Epistemological Foundations of the Free Society

Also by Jerry Kirkpatrick

In Defense of Advertising:
Arguments from Reason, Ethical Egoism,
and Laissez-Faire Capitalism

Montessori, Dewey, and Capitalism:
Educational Theory for a
Free Market in Education

Independent Judgment and Introspection:
Fundamental Requirements
of the Free Society

Applying Principles: Short Essays
Based on the Philosophy of Ayn Rand,
Economics of Ludwig von Mises,
and Psychology of Edith Packer

Epistemological Foundations of the Free Society

Ayn Rand's Theory of Universals
Applied to Science, Sigmund Freud's Psychology,
and Ludwig von Mises's Economics

JERRY KIRKPATRICK

KIRKPATRICK BOOKS
UPLAND, CALIFORNIA

Publisher's Cataloging-in-Publication Data

Names: Kirkpatrick, Jerry, author.
Title: Epistemological foundations of the free society : Ayn Rand's theory of universals applied to science , Sigmund Freud's psychology , and Ludwig von Mises's economics / Jerry Kirkpatrick.
Description: Includes bibliographical references and index. | Upland, CA: Kirkpatrick Books, 2026.
Identifiers: LCCN: 2026907179 | ISBN: 978-1-7371387-5-4 (hardcover) | 978-1-7371387-6-1 (paperback) | 978-1-7371387-7-8 (ebook)
Subjects: LCSH Rand, Ayn--Philosophy. | Rand, Ayn--Criticism and interpretation. | Objectivism (Philosophy) | Freud, Sigmund, 1856-1939. | Von Mises, Ludwig, 1881-1973. | BISAC PHILOSOPHY / Epistemology | PSYCHOLOGY / General | BUSINESS & ECONOMICS / economics / General | PSYCHOLOGY / Psychotherapy / Psychoanalysis
Classification: LCC B945.R234 K49 2026 | DDC 191--dc23

Library of Congress Catalog Card Number: 2026907179

Cover design by 1106Design, 1106design.com

Kirkpatrick Books, Upland, CA 91784
http://www.jkirkpatrick.net, jkirkpa380@gmail.com

Contents

II. Freud and Mises

III. Final Thoughts

Preface

A sound theory of universals is fundamental to the defense of a free society.

Plato said that universals, which he called forms, are located in another dimension of reality and only certain people can know them. Hence, his advocacy of a near-communistic society of philosopher-kings who would rule. Thomas Hobbes advocated a theory called nominalism, which says universals are conventional, arbitrary names, not sound or objective concepts that are true. This made him a relativist in ethics and advocate of a strong "public sword," or monarchy, to keep the peace in society.

Karl Marx did not have a theory of universals, though we might say it was implied in what Ludwig von Mises calls Marx's polylogism, a theory of many logics that rejects the notion of only one fundamental logic based on Aristotle's three laws. Thus, Marx claimed that the bourgeoisie have their logic, while the proletariat have theirs, which means the two classes cannot talk to or understand each other. Hence, his advocacy of the dictatorship of the proletariat to establish that Garden of Eden called communism—or socialism—he considered both terms synonymous. Today's postmodern Marxists say something similar: "You have your truth; I have my truth." As consequence, what they advocate in political philosophy is a repressive society, with the so-called victim classes taking the place of the proletariat.

Underlying any theory of political philosophy is a theory of human nature, also called philosophical psychology. A sound theory of universals is required to form and define the concepts that identify and define the essentials of human nature. Do human beings possess the *capacity* to reason, the word "capacity" meaning we do not reason automatically but must choose to exercise that capacity, which means we have an active consciousness with which we control our lives? Or are we determined by genes and environment with illusions of consciousness and free will? The illusion of consciousness is called materialism and the illusion of free will is called determinism. Both are advocated by Hobbes and Marx. If we do possess a volitional consciousness, however, the idea of a free society—capitalism—follows from basic concepts of rights, freedom, the moral validity of the pursuit of self-interest, and, in particular, a sound theory of universals that explains how objectivity and truth are possible.

The significance of a correct theory of universals is that our knowledge, with the exception of proper names, is entirely in the form of universal concepts. If those concepts are not valid, they mean nothing when we move from epistemology to ethics and politics, and to all the special sciences, including psychology and economics. That is, we must be able to use and apply the theory of universals to validate science, especially our concept of human nature—to establish the foundations of psychology, to validate the concept of value to establish a theory of ethics, and to validate many other concepts to define the proper function of government. This last includes rights and freedom, along with such additional concepts as the nature and meaning of healthy and unhealthy motivation and behavior and the production of wealth in a social setting.

A correct theory of universals means our minds are competent to know reality and thus can guide us in our choices and actions to grow, thrive, and prosper.

Ayn Rand's theory of concepts, which I will argue is a sound theory of universals, is her most important and significant contribution to philosophy, especially to epistemology. Only by making the theory widely known and understood can the free society be properly defended.

The present work is not a defense of capitalism per se. Ayn Rand and Ludwig von Mises provide unanswerable moral and economic justifications.[1] The subject of this book is epistemology and its importance in identifying the nature of science, basic and applied, with emphasis on the human sciences of psychology and economics. Its theme is

---

[1] I must add George Reisman and his magnum opus *Capitalism: A Treatise on Economics*. Reisman is a follower of Ayn Rand's philosophy and former student of Mises.

that the essence of science is observation, not exact measurement, and that science employs as its primary and fundamental method the mental process of conceptualization: generalization from observation of particulars to identify universals, then application of previously formed universals to understand new particular cases. The issue of universals, based on Ayn Rand's theory of concepts, runs throughout, providing the fundamental epistemological foundation for proper defenses of reason in philosophy, individualism in psychology, and capitalism in economics.

My audience, as a result, is laypersons and academics interested in the underlying epistemology of science, especially the human sciences that, when integrated with psychology and economics, provide the ultimate vindication of a free society and answer to such ill-founded theories as those of Plato, Hobbes, and Marx.

In chapter 1, using Rand's theory of universals, or concepts as she calls them, I introduce her notion of measurement omission as the essence of abstraction to address the proper meanings of, and distinctions between, basic and applied science, theory and history, and the cognitive processes of generalization and application. Positivism—logical positivism or logical empiricism, as it is also labeled—and experimentation are discussed, followed by a presentation of what is called the problem of universals.

In chapter 2, I elaborate Rand's theory of concepts as an improvement on Aristotle's theory of form and matter by identifying concepts and essences as neither intrinsic, in the thing, as Aristotle taught, nor subjective as the nominalists assert. Rather, concepts and essences are epistemological, in our minds, created by us and are valid and epistemologically objective if they correctly identify the metaphysical referents on which they are based. I further discuss Rand's important explanations of abstraction from abstractions and concepts of consciousness. In concluding the chapter, I elaborate Rand's usage of the terms "metaphysical" and "epistemological," ending with a presentation of what I consider her radical theory of logic as non-contradictory identification.

In chapter 3, I discuss several related philosophical issues using Rand's theory to refute the bane of science, positivism, and to provide valid empirically-based explanations of Immanuel Kant's allegedly innate "a priori" concepts—space, time, and causality—as well as a non-religious, non-Kantian empirical defense of free will. I continue with a discussion of the distinction between general knowledge, which we all possess to some degree and with some accuracy, and personal knowledge, which is unique to our own experiences. Finally, I conclude with a brief discussion of the role and importance of emotion in both science and our personal lives.

In chapters 4 and 5, respectively, I apply Rand's theory to my understanding of psychology and economics as human sciences, with emphasis on the work of Sigmund Freud and Ludwig von Mises. Both Freud and Mises have been criticized as pseudo- or non-scientific for using "literary" (Freud) or "non-quantitative" (Mises) methods to establish their theories. To the contrary, I argue that the methods of both are models of conceptualization, which I contend is the fundamental method of science.

I became aware of the importance of conceptualization in science when I reread Edith Packer's *Lectures on Psychology* after they had been made available as a Kindle book, in 2013, by her husband George Reisman. I realized after reading the book that conceptualization was what Dr. Packer was doing as a psychotherapist, and then realized that Mises, Dr. Reisman, Aristotle, and, when reading him later, Freud, were doing the same thing in their work. Masses of quantitative data and laboratory experiments, I realized, were not the essence of science.

In chapter 6, I conclude the work with a summary of its thesis, elaborating on what I call thinking in definitions and clarifying the meanings of scientific law and principle. Finally, statements about Kant's legacy and, subsequently, Ayn Rand's complete the book.

The significance of focusing on Freud and Mises is that they both provide fundamental concepts and principles for the foundation of a free society, Freud for the individual to pursue a healthy, happy, and psychologically independent life and Mises for individuals to cooperate with each other to achieve peace and prosperity. Both are experts on human nature, explicitly so for Freud, implicitly for Mises, and both, despite the comments of their critics, consider themselves to be scientists. This work aims to provide Freud and Mises a needed epistemological foundation.

The meaning of free society that I am using is Ayn Rand's concept of laissez-faire capitalism as a *social* system, presupposing an ethics of rational self-interest. That is, capitalism is not just the private ownership of the means of production, as economists, Mises in particular, define it, but a system based on the primacy of individual rights, especially property rights, with all property privately owned, allowing social cooperation to be achieved by voluntarily trading value for value. Rights are freedoms of action—deriving from human nature as a rational being—to speak and write openly and to work with others to pursue property, careers, relationships, hobbies, education, and happiness without the interference of *initiated* physical coercion, that is, physical force, by the government or criminals. Self-defensive coercion is delegated to the government unless involved in an emergency situation.

The importance of defining capitalism as a *social* system is that it broadens the concept beyond economics alone. This is significant because totalitarian states are called such because they seek to and often do—in the name of an ethics of self-sacrifice—control every aspect of the individuals' lives, not just their economic lives.[2] Totalitarianism means *total control*. Capitalism means *total freedom*—from direct or indirect initiated physical force of any type, economic or political.

A further point about the meaning of Rand's concept of capitalism. It is not what we have today or had in the late nineteenth century, though Rand acknowledges the United States came close in those earlier years. Her concept is a goal, a moral and political ideal, to work toward to achieve at some point in the future. It is not Marx's utopia or creation of a "new man," as Marx said communism would give us. Capitalism requires hard work and self-responsibility with no one expected to provide handouts. The problems with the apparent capitalism that we had in earlier days, and have today, is that many big businesses and governments have colluded with each other to gain favors and wealth for themselves, ignoring everyone else. This is not capitalism, but corporatism or, more precisely, fascism.

For this reason, one of the best ways to describe capitalism is that it is a system in which business and state are completely separated in the same way and for the same reasons as we now have the (nearly) complete separation of church and state. And by "business," I mean any organization that operates to earn a profit by offering value to paying customers. I include schools, roads, and utilities, today's publicly owned and mixed institutions. I also believe that there should be no so-called nonprofit organizations. All societal institutions, business, church, or otherwise, should be privately owned and operated and expected to earn an excess of revenues or donations over expenses. If there is to be taxation to support the considerably limited government, it should be a

[2] The ethics, especially today, is usually a variant of Kant's duty-over-inclination maxim, imposed at the point of a government gun (or Hobbes's public sword). The individual's life belongs to the state and must be sacrificed to it when and as requested.

flat tax applied to everyone, including to those so-called nonprofits. As Ayn Rand indicates, there are other options in financing government in a free society, but the issue is far removed from today's problems. For the most part, it should be postponed for debate until we come closer to a truly free society.

I did not come up with the theme of this book until recently, but the idea behind it was implanted in me during the spring quarter of my freshman year, 1966, at the University of Denver. I took the course "Modern Philosophy," taught by Leonard Peikoff. The course covered Thomas Hobbes, the Continental rationalists (Descartes, Spinoza, Leibnitz), the British empiricists (Locke, Berkeley, Hume), and Immanuel Kant—in ten weeks.[3] The issue of universals and political philosophy arose immediately because Dr. Peikoff's lectures began with Hobbes and the consequences of Hobbes's nominalism. The issue continued throughout the course, at least indirectly, when we came to John Locke's inability to find a "something I know not what," Hume's failure to find a necessary connection between cause and effect, and, of course, Kant's failure to find reality.

An important note here is that Ayn Rand's theory of concepts did not begin to be published until July of that year, and Dr. Peikoff's teaching style was to present the best case he could for a viewpoint he did not accept, even bending over backwards to make it sound like he supported it. He knew Ayn Rand's theory at this time but did not present it in the course.[4] Lively Q&A's went on throughout the term, but no solution to the problem of universals was found. I concluded, reluctantly, that those of us most interested in the topic were left on the floor, desperately trying to find Aristotle's intrinsic essences. After that quarter, Dr. Peikoff returned to New York City where he had lived for many years.

---

[3] The quarter system at the time was ten weeks plus a week of finals, and classes were five days a week, fifty minutes each. I took Professor Peikoff's introductory philosophy course during the previous winter quarter, along with 160 of my closest classmates. The modern philosophy class consisted of about thirty-five students.

[4] In the introduction to philosophy course, Dr. Peikoff had nearly everyone convinced he was a determinist. In the modern philosophy course, I knew that he knew Ayn Rand's theory because he taught it in a graduate seminar during the fall quarter of 1965. I even had the audacity to walk into his office, before attending a single college class, to ask him if I could audit his graduate seminar. He said "no," because it would not be fair to others. He did, however, give me the telephone number of the local representative of the Nathaniel Branden Institute that presented lecture courses on tape of Ayn Rand's philosophy.

When Ayn Rand's theory was published, I memorized key definitions to think about them, for example, when waiting for my next class.[5] In the ensuing years, my interest in epistemology and the problem of universals never waned. It was a treat to read the discussions of epistemology by Ludwig von Mises and, though not exactly a treat, my knowledge of epistemology helped me in graduate school to fend off the assaults of that god called positivism.

My relationship to Ayn Rand's philosophy, which she named Objectivism, is that I am a follower of her philosophy, but do not describe myself as an Objectivist. I call myself a student of Ayn Rand, in the same way I call myself a student of Sigmund Freud and Ludwig von Mises. In 1968, in the aftermath of Rand's break with Nathaniel Branden, she emphasized that Objectivism was her philosophy and that followers should take care not to speak for her, but to continue working as independent intellectuals in their own fields applying her ideas as relevant.[6] This is how I have viewed my scholarly work for many decades. In the present book, I emphasize that I am presenting and applying her ideas as I understand them. Elaborations and applications are mine.[7]

My thirty-six-year academic career was spent in the applied science of business marketing, teaching and writing about fundamental ideas—Ayn Rand's, as well as those of psychologists and economists—applying them to the business disciplines and education.

The many references used in this book indicate a wide variety of influences on my thinking and writing. I endorse some, reject others or attempt to refute them, and some are neutral, providing important information as background and history. Aside from Ayn Rand, Freud, and Mises, my most significant influences are psychologist Edith Packer

---

[5] Rand's theory of concepts was initially published in eight monthly installments in *The Objectivist* magazine from July 1966 to February 1967. One of the (non-epistemological) definitions I memorized was her notion of capitalism as a social system. When I mentioned it to a classmate, the classmate was surprised, and I was surprised that he was surprised—because I had assumed that Rand's definition was self-evident and that everyone must have known it!

[6] Ayn Rand, "A Statement of Policy," *The Objectivist* 7, no. 6 (June 1968): 7–9.

[7] In the present work, I have concentrated on avoiding any tinges of moralism, a trait that unfortunately is sometimes present in Rand's writing and in the demeanor of some of her followers. See my discussion of the anti-conceptual mentality in Jerry Kirkpatrick, *Independent Judgment and Introspection: Fundamental Requirements of the Free Society* (Kirkpatrick Books, 2019), 108–110, and below, chap. 3, the section titled "Defensive Habits and Free Will," pp. 67–69.

and economist George Reisman. My primary and permanent influence is my wife, soulmate, and editor, philosopher Linda Reardan. Her conceptions both of post-Kantian, modern Aristotelianism and of emotions as data for reasoning have influenced this book. Our daughter, a mechanical engineer, has also been helpful when challenging me to explain to her, in simple terms, the thesis of the work.

The formatting, as with all my books, follows *The Chicago Manual of Style*, though not completely. I still prefer superscript numbers in the footnote reference, for example, as opposed to Chicago's recommendation to use full-sized numbers followed by a period. Clarity and consistency are the criteria of formatting, which I learned from Carol Fisher Saller, *The Subversive Copy Editor.*

For references to the internet, I have shortened most of the uniform resource locators (URLs) to their domain names. When searching for my reference, I recommend using what is called the "site operator," for example, *site:freud.org.uk private meaning* or *site:mises.org economic calculation.* This will almost always, at the top of the search, return the words that follow the domain name. Digital object identifiers (DOIs), unique locators usually of academic journal articles, have not been shortened. Citations to print publications have been shortened after the initial reference. All sources are listed in the bibliography in full, including uniform resource locators.

A final note on "self-plagiarism," a term I believe to be a ludicrous contradiction in terms that has gained some notoriety in the academic world. I emphatically do not endorse the notion that a writer can plagiarize him- or herself, especially when acknowledgment is made about the so-called recycling of one's own work. For the record, I have been writing and posting monthly blogs for nineteen years and more than twenty of the posts, since 2021, were written as drafts for eventual use, after editing, in the present work. A few excerpts from the draft of this work have also been posted as blogs.[8]

[8] See jerrykirkpatrick.blogspot.com for the blogs. I have always considered my blog posts to be drafts of whatever I might later publish. See Richard Posner, *The Little Book of Plagiarism* (Pantheon, 2007), for his discussion of the legal and moral issues involved in plagiarism and "self-plagiarism." The term "self-plagiarism" likely applies more correctly to students who turn in the same paper for two different classes.

I. Universals

1

The Essence of Science

> The process of observing the facts of reality and of integrating them into concepts is, in essence, a process of induction. The process of subsuming new instances under a known concept is, in essence, a process of deduction.
>
> —Ayn Rand[1]

> Thus demonstration develops from universals, induction from particulars. . . . It is sense-perception alone which is adequate for grasping the particulars.
>
> —Aristotle[2]

THE PURPOSE OF SCIENCE is to explain and guide.

The fundamental method scientists use is observation to raise questions followed by frequent testing and trying of alternatives to find correct answers. The mental process is conceptualization, which consists of generalization to identify universals and application of previously formed universals to understand particular cases. The former is a process of induction, the latter of deduction. The two together constitute what is called scientific method.[3]

[1] Ayn Rand, *Introduction to Objectivist Epistemology* (NAL Books, 1990), 28. First published in *The Objectivist* 5, nos. 5–12; 6, nos. 1–2 (1966–67). Expanded 2nd ed., with "The Analytical-Synthetic Dichotomy," by Leonard Peikoff, and excerpts from epistemology workshops, 1969–71.

[2] Aristotle, *Posterior Analytics*, 81a40, 81b8, in Richard McKeon, ed., *The Basic Works of Aristotle* (Random House, 1941), 136.

[3] See Aristotle on induction and deduction, *Posterior Analytics*, 71a–72b4, 81a39–81b9, 99b15–100b18, in McKeon, *Basic Works*, 110–13, 135–36, 184–86.

All of us—laypersons and scientists—use both cognitive processes, but scientists use them more intensely and in more concentrated areas.

Scientific method may or may not involve controlled experimentation with an elaborate array of instruments and it may or may not produce mathematical equations and hosts of quantitative data.[4]

The aim of science is not ivory-towered contemplation for its own sake disconnected from reality. Its purpose is practical, to enhance human life, which can only be fully accomplished in a free society with private property and freedom of action, which includes free inquiry and debate, which means capitalism.

THE SEARCH FOR UNIVERSALS

The 1981 book *Great Scientific Experiments* by Rom Harré presents in popular format and in only 200 pages the gist of twenty influential scientific experiments, from Aristotle to the twentieth century. Interestingly, most of the experiments cited do not use statistical samples of any size divided into experimental and control groups. Most samples are small and one is a sample of one. How is this possible? And if true, may we conclude that such scientific outcasts as Sigmund Freud, Jane Goodall, and Ludwig von Mises were great experimenters?

Science and Positivism

It is the positivism of modern science that tells us a sound theory of induction does not exist, because universals do not exist, only statistical probabilities and correlations. Therefore, anything close to causal must be found through at least two groups of large samples to control possible confounding variables or, what is stated more formally, sources of extraneous variation. Hence, all we can find through our many studies are "successive approximations." This is what John Stuart Mill's followers called the hypothetico-deductive method that drives nearly all research today in both basic and applied science and condemns Freud, Goodall, and Mises to the realm of "pseudoscience."[5]

---

[4] Cf. Armand Marie Leroi, *The Lagoon: How Aristotle Invented Science* (Penguin, 2015), for a summary of Aristotle's biological science.

[5] Steffen Ducheyne and John P. McCaskey, "The Sources of Mill's Views of Ratiocination and Induction," in *Mill's A System of Logic: Critical Appraisals*, ed. Antis Loizides (Routledge, 2014), 76–77. Mill more simply called the method deductive or hypothetical.

Two key terms need to be defined. Science, the first, is a systematic study of an aspect of reality that explains its domain descriptively and causally, and if applied (as technology) provides guidance for human choices and actions to achieve specific goals. The product of any science is a body of knowledge—a collection of integrated, universal concepts and principles based on relevant facts.

A second term to look at is experimentation. The Latin root of the verb "to experiment" means to try. Paraphrasing the *Oxford English Dictionary*, to experiment means to test or try something to identify what has previously been unknown.[6] This means experimentation is trial and error. It may involve experimental and control groups (and the manipulation of one variable to determine its effects on another), but it does not necessarily have to. It also may involve the use of laboratory apparatus, or it may be entirely conceptual—in the head, sometimes called a thought experiment.

Indeed, formal experimental methods likely derived as an analogy from introspection into how we test and try to come up with correct concepts. Socrates is the leading example of this early experimentation when he asked his fellow Athenians to identify the correct meaning of such concepts as piety, justice, courage, and beauty. In the process, Socrates set aside the nonessential notions proposed by his counterparts as useless or sidetracks to the correct identification. Setting aside useless and irrelevant or nonessential ideas, we might say today, was Socrates's way of holding constant or eliminating possible confounding variables or sources of extraneous variation. Today, we have more sophisticated methods, at least in certain fields of research, of holding constant those confounding variables.

Science historian John P. McCaskey has traced the history of induction and, by implication, the role of experimentation in it. McCaskey found that until the last couple of hundred years induction and experimentation were understood to be based on the work, among others, of Francis Bacon, who in turn based his work on Aristotle's formal cause. Aristotle in turn developed his theory of induction from what during the fifteenth and sixteenth centuries was called "Socratic induction."[7]

---

[6] *Oxford English Dictionary*, s.v. "experiment (v.)," June 2024, https://doi.org/10.1093/OED/6224179652.

[7] John P. McCaskey, "Induction in the Socratic Tradition," in *Shifting the Paradigm: Alternative Perspectives on Induction*, ed. Louis F. Groarke & Paolo C. Biondi (De Gruyter, 2014), 161–92, https://doi.org/10.1515/9783110347777.toc. McCaskey, "Induction, Philosophical Conceptions of," in M. Sgarbi, ed., *Encyclopedia of Renaissance Philosophy* (Springer, 2020), https://doi.org/10.1007/978-3-319-02848-4_1054-2. The Socratic and Aristotelian methods

Socrates, McCaskey points out, was looking for universal essences that apply in all instances. Universals, those exceptionless concepts and principles, not probabilities or correlations, are what constitute the essence and foundation of science.

The Fundamental Method of Science

Harré's historical survey of experimentation shows that conceptualization, or concept formation (and application), is the fundamental method of science, because large samples are often not necessary to identify the universals. Harré writes:

> It was [Francis] Bacon who first saw clearly that experiments must serve the complementary functions of suggesting definitions of the nature of things and of eliminating those that are useless . . . Bacon realized that the aim of experimental science is the refinement of our ideas about the natures or essences of the substances, properties and processes we find in the natural world. Typical scientific questions would be "What is colour?" "What is liquefaction? "What is heat?" In answering such questions, we would have to formulate definitions of the nature of these things, processes, properties and so on.[8]

McCaskey, as do I, acknowledges Ayn Rand as providing the theory of concept formation and its implied role in science, specifically the inductive process of identifying essential distinguishing characteristics.[9]

Here are three examples from McCaskey of scientific conceptualization. Robert Koch identified what he called the comma bacillus, because it looked like a comma, as the defining, universal characteristic and cause of cholera, not such competing but correlational hypotheses as season of year or foul water. Charles Wells concluded, after many small experiments to test a wide variety of independent variables, that water condensation was the essence of dew. And Lord Kelvin, despite

---

induce from particular *things* or *concretes*, whereas the Scholastics and today's Humean and Millian followers induce from particular *statements* or *propositions*, often making the entire process rationalistic and cut off from reality.

[8] Rom Harré, *Great Scientific Experiments: Twenty Experiments that Changed Our View of the World* (1981; repr., Dover Publications, 2002), 6, 13.

[9] McCaskey, "Induction in the Socratic Tradition," 187.

two centuries of the collection of quantitative data, reasoned that tides by definition, following Newton, are "motions of water on the earth, due to the attractions of the sun and of the moon."[10]

In the study of tides there likely was confusion, as happens in developing sciences, over the difference between basic and applied science or technology. Sea captains need to know the timing and heights of tides at a specific location, and statistics can help make those predictions, hence the many early studies to collect data on tides at various ports. But this is applied science where the universals of theory are used to uncover correlations, tolerances, and averages to guide concrete, practical decision making.

Similarly, the unique, individual planets Uranus and Neptune were discovered through application of accumulated knowledge of the night sky. Newton's laws in particular were used to identify a tiny segment to enable astronomers' telescopes to find the "little wanderer" that became the newly known planet Neptune. The mathematics were extensive, but the science was applied, not basic.[11]

Harré acknowledges that one or a few cases, with careful experimentation, that is, trial and error, can yield "defining properties of all samples similar to them." He calls the small samples "intensive design." In contrast, he calls the larger versions "extensive."[12]

Harré recounts an intensive design, with a sample of one, in his second chapter: the nine years of experimentation conducted by army doctor William Beaumont on Alexis St. Martin, whose stomach in 1822 did not close completely after a musket wound. Access to St. Martin's stomach contents, especially his gastric juices,

---

[10] Kelvin is quoted in McCaskey, "Induction in the Socratic Tradition," 186. McCaskey also presents Ohm's law as another example, but I have not included it here. See, in addition, three blog posts at *John P. McCaskey* (blog), johnmccaskey.com: "Key to Induction: Distinguish General and Universal," June 11, 2014; "The History of Induction," March 12, 2015; and "Induction without the Uniformity Principle," May 12, 2015. On the usage "by definition," see chap. 6 below, p. 156.

[11] Laura J. Snyder, *The Philosophical Breakfast Club* (Broadway Books, 2011), chap. 11. The Greek word *planetai* means wanderers. Since ancient times, planets have been referred to as "little wanderers" or "wandering stars." They were called this because their paths across the sky constantly change in relation to the paths of the "fixed" stars.

[12] Harré, *Great Scientific Experiments*, 193. Harré sometimes uses the words "samples of substances" to refer to a few or several individual cases of, say, an element, disease, or natural process to arrive at a precise identity, that is, its defining properties.

enabled Beaumont to answer the question, are gastric juices chemical solvents or is the process of digestion some vital force and the juices just inert water?

Feeding different foods to his subject to observe digestive processes and removing gastric juice to test its effects in a glass jar were the experiments that Beaumont conducted to draw the conclusion of chemical solvency, even of "the hardest bone." The concept correctly identified by Beaumont was that gastric juices were—and universally so—chemical solvents, not inert water.[13]

In my book *Independent Judgment and Introspection,* I praise Sigmund Freud for his work in clarifying and defining psychological repression and Jane Goodall for her discoveries about chimpanzees. Both conducted great experiments, the former in psychology, the latter in primatology, a branch of biology.[14] They both used the method of conceptualization and deserve to be recognized as great experimenters. Science, in other words, is not restricted to the fields of physics and chemistry, or imitations of those fields, as positivists are prone to assume.

Freud spent over thirty years in talk therapy with many different patients before he finally formulated the essential meaning of repression as an unconscious (or subconscious) response to anxiety that mutes or entirely blocks the experience of thoughts, memories, or emotions from conscious awareness. He also had to distinguish repression from defense mechanism in general, after tending to equate the two for many years.[15]

Repression is an aspect of all defense mechanisms, or defensive habits, as I prefer to call them, but it is not identical to them. Freud's daughter Anna said that it was in his 1926 book *Inhibition, Symptom, and Fear* when Freud identified the correct place of repression in psychology.[16] The muting or barring of an emotion is accomplished by muting or barring the thought that stands behind it.[17]

---

[13] Harré, 41. Beaumont became famous because of his work with St. Martin.

[14] Kirkpatrick, *Independent Judgment*, 78–79. The master of thought experiments was Ludwig von Mises. See chap. 5 of the present work.

[15] Kirkpatrick, 79n51.

[16] Anna Freud, *The Ego and the Mechanisms of Defence*, trans. Cecil Baines (International Universities Press, 1946), 45–46. First published in German in 1936. See below, chap. 4, p. 81n13, for the justification of using "fear," instead of the more recognizable "anxiety," in Freud's title.

[17] Kirkpatrick, *Independent Judgment*, 145n11. I prefer the words "defensive habits" to avoid any hint of mechanistic materialism, because defenses are mental habits that we form and adopt, usually when young. Traumatic memories,

Jane Goodall's work, beginning in 1960, was to observe the behavior of chimpanzees in the Gombe Stream Chimpanzee Reserve (now the Gombe National Park) in Tanzania. Over the course of many years, she discovered that chimps eat meat, make elementary tools to obtain food, can be violent and even cannibalistic, and have personalities. She gave the chimps names and interacted with them after she had gained their trust.[18]

Goodall's work required patient observation and, especially, application of analogous concepts from human psychology to the higher mammals, "patient observation" meaning testing and trying different ideas before drawing conclusions.

It is the positivists who dismiss Freud's work as anecdotal at best, literature at worst. And it was the "men of hard science" at the beginning of Goodall's career who dismissed her as an uncredentialed woman frivolously and anthropomorphically giving names to the nonrational, dumb animals. Paleontologist Stephen Jay Gould, however, calls Jane Goodall's work "one of the Western world's great scientific achievements."[19]

It is the positivists and many of those "men of hard science" who in fact fail to understand the nature of science and great experiments.[20] They relegate values to the netherworld of subjectivism and consciousness and free will to the wasteland of the metaphysically meaningless. The notion of universals and the so-called problem of universals are rarely discussed today—because what the positivists have taught us over the years is that probabilities and correlations are "good enough." This has been called "complacent skepticism."

I have suggested, are not repressed and should not be thought of as part of the definition of repression. "Subconscious" is today's more preferred term than "unconscious."

[18] Jane Goodall, *In the Shadow of Man* (1971; Mariner Books, 2000), esp. chap. 3 and 6.

[19] Quoted in Goodall, *Shadow of Man*, xvii. Mises, of course, is criticized for being non-quantitative and passé.

[20] Robert Whitaker, *Mad in America: Bad Science, Bad Medicine, and the Enduring Mistreatment of the Mentally Ill* (Basic Books, 2002), 30–38. The words "men of hard science" are Whittaker's, mocking the materialistic determinists of the latter half of the nineteenth century who denigrated as old-fashioned, religious "gardeners and farmers" the moral treatment movement of the Quakers. Whittaker, 37. The movement emphasized kindness, attention, listening, and talking to help the mentally ill. Unfortunately, the movement died out by 1890. I have extended the use of Whittaker's words to positivists and the "science is measurement" crowd who, when they dislike or disapprove of something, unfortunately often label it *soft* science or *not* science at all.

THE SCIENCES—BASIC AND APPLIED

Let me now set the broader context of the different sciences and what distinguishes them before getting into the problem of universals.

The most fundamental science is philosophy or, as some have said, philosophy is the science of all sciences, with epistemology providing the principles of method to guide scientists—and laypersons—in their choices and actions.[21]

The Special Sciences

There are three fundamental (basic) special sciences: the physical that study inanimate matter and its motions; the biological that study living organisms and their behavior; and the human that study homo sapiens—the highest living organism that possesses a consciousness with the capacity to reason—and its actions.[22]

Special sciences representative of each, respectively, are physics and chemistry, botany and zoology, psychology and economics. All other sciences derive their basic premises from one or more of the fundamental ones. Derivative sciences are more applied than the fundamental ones and at some point may be called technology or applied science—engineering, medicine, psychotherapy.

Applied sciences are largely, though not exclusively, deductive and prescriptive. The fundamental sciences are largely, though not exclusively, inductive and descriptive.[23]

---

[21] A character in Jack London's 1908 novel *The Iron Heel* (1908; repr. Penguin Books, 2006), 16, says, "Philosophy is merely the widest science of all. . . . the science of science, the master science," and attributes a similar conviction to Herbert Spencer. Logic for centuries has been called the science of sciences, *scientia scientiarum*. The premise is ancient, attributed to Aristotle, as the special sciences over time separated from philosophy to concentrate on their unique areas of existence. A cursory internet search reveals that the phrase today, unfortunately, is also associated with mysticism, as in "spirituality is the science of all sciences." The words "science of science" are also used to study the foundations of physical science. Cf. H. W. B. Joseph, *An Introduction to Logic* (Oxford University Press, 1916), 3.

[22] In an earlier era, I would have used the word "man" for homo sapiens, which I still believe is the more correct and concise term.

[23] Throughout this and the next chapter, I focus primarily on the formation of concepts through generalization. Much of basic science, however, also requires the skill of application, for example, to narrow a search to a key area for further investigation or to a key variable, such as Koch's comma bacillus to test whether it is truly universal. Science requires the skill of frequently alternating between induction and deduction.

To return to the first sentence of this chapter, the aim of science is not explanation and prediction as the positivists insist, but explanation and guidance.[24] To explain means to describe the nature of an entity and its attributes and actions, which means explanation includes the ability to predict. To guide means to provide principles of action to help human beings take necessary steps to achieve their chosen goals.

Principles of guidance are everywhere, used in all areas of our lives. They can be used to build spaceships and bridges; to diagnose diseases in plants, animals, and humans to recommend treatments; to produce wealth and prosperity; to manage small and large businesses; and to decide whether a specific action would be honest. They can also be used to weave baskets and to help a child make a toy boat to float on the local pond or swimming pool.[25]

Testing and trying are what anyone with the capacity to reason does to conclude that a concept's essential distinguishing characteristic has been identified. It is important to emphasize that when acquiring new knowledge laypersons perform exactly the same mental processes, in varying degrees of accuracy, as laboratory scientists and ivory-towered academics. The aim of all learning is to improve the well-being and enjoyment of our lives.

A child, for example, may throw, bounce, and chew a round, spongy thing before concluding that the round, spongy thing is a ball that bounces and rolls—but is not food. Aristotle opened twenty fertilized chicken eggs, one per day, to observe the development of chicks from beating speck of blood to live baby chicken. In this way, he tested and confirmed the hypothesis that baby birds are not born pre-formed but grow gradually stage by stage within the egg.[26] The improvement of life deriving from both examples? A ball is something that can now be played with, giving the child considerable pleasure, and the farmer raising chickens can now exert more effort incubating the fertilized eggs or, rather, allowing the hens to incubate without interruption.

---

[24] Ayn Rand, "Philosophical Detection," in *Philosophy: Who Needs It* (Bobbs-Merrill, 1982), 17. First published in *The Ayn Rand Letter* 3, nos. 9–10 (1974).

[25] The literature on basket weaving, for both beginners and the advanced, is substantial. Ethics is a fundamental branch of philosophy that provides abstract principles of guidance, leading us throughout our lives to make choices and take actions—honest ones, for example—consonant with our human nature that possesses a volitional, rational consciousness. Ethics teaches us how to live the good—meaning moral—life.

[26] Harré, *Great Scientific Experiments*, chap. 1.

Today's scientists test and try many options, sometimes using controlled experimentation, to eliminate sources of extraneous variation. In the physical sciences, they may identify algebraic equations to explain the relationship between two or more entities or their attributes or actions.

The biological and human sciences, however, because of the nature of their subjects cannot be so precise. Concept formation, as Ayn Rand has shown, involves measurement—and measurement omission to arrive at the universal—but these sciences can only be quantified to a limited extent, using approximate measurements, such as greater than or less than, higher or lower, more important or less important.[27]

The human sciences have the added proviso that free will reduces to absurdity the formulation of algebraic equations to predict behavior. Gestalt psychologist Kurt Lewin proposed this formula to explain human behavior: $b = f(p, e)$, where b is behavior that is supposedly explained and predicted by the interactions of person variables, p, and the environment (or situation), e.[28] Subsequent debate focused on whether the two variables were additive or multiplicative. The two variables are not irrelevant in explaining human behavior, but absent any discussion of free will, the formula is as absurd as many others, such as $q = 2000 - 2.5p$ of the mathematical economists, where q is quantity demanded that is allegedly explained and predicted by price, p.

Determinism is the assumption underlying the use of exact equations in the theory of any basic human science. Should we not add a third variable to Lewin's equation, v for volition, and emphasize that it could be a constraint on the influence of p and e? Or is it p and e that constrains v? Equations (and statistical studies) may be helpful, if well-constructed, in the applied human sciences, for example, requiring forecasting. The law of supply and demand in economics can be expressed as an equation, $p = d/s$, but only for general, inexact predictions. The data used for such forecasting is history, not theory.[29]

---

[27] Rand, Objectiv*ist Epistemology*, chap. 1 and 2.

[28] *Principles of Topological Psychology* (McGraw-Hill, 1936), 11–13.

[29] Weather forecasting, for example, applies the universal principles of physics to accumulated historical data to predict weather at a particular location and time. Forecasting extrapolates from the recent past to predict the near future. It may or may not be aided by universal principles and elaborate mathematical models.

All sciences work with universals by omitting—but not forgetting—the measurements. The essence of science is conceptualization, which requires measurement omission, not measurement. Algebraic equations in the physical sciences, $f = ma$ or $e = mc^2$, for example, omit the range of quantities of each of the equation's variables (but not any constants, of course).

Essentialization in the formation of concepts is the fundamental thinking method of everyone, scientists and laypersons, because the essential distinguishing characteristic, or essence, of a concept, identifies what explains and causes all or most of the other characteristics. The flat, level surface of a table explains why my glass of water sits stably on it and, along with the essential characteristics of the glass—cylindrical shape with a closed bottom—causes the liquid to stay within it. The motions of the sun and moon explain and cause the tidal motions of water on earth. Essentialization is a process of thinking in causes and effects.

Measurement is important in all sciences, but especially so in technology or applied science, to quantify individual cases. The engineer needs a variety of measurements to determine the best location and structure of a new bridge, including the width of the river and the nature and history of its soil. The sea captain requires reliable estimates of the tides in specific harbors, the medical doctor measures the patient's temperature and blood pressure, and the psychotherapist estimates the kind and quantity of thinking errors an unhappy patient has made over the years.

Principles of guidance in applied science explain the nature of the goals one is seeking and the steps that will cause or produce that accomplishment. Most applied sciences, including some fundamental ones, such as epistemology and ethics, are "how-to" disciplines. To sell a product, for example, the salesperson must tailor his or her presentation to the prospects' requirements for the improvement of their lives (their needs) and to their optional desires or tastes (their wants). To weave a basket, one must collect the appropriate materials, then follow required steps to lace strands of material over and under each other to create the desired shape. In epistemology, the goal is to identify how we acquire new knowledge, so we follow the principles of logic, especially Aristotle's law of non-contradiction, to ensure that the contents of our consciousness do not contradict the facts of reality.

Theory and History

Where does history fit? Throughout the nineteenth century, and still today, historians, along with psychotherapists, medical doctors, and most anyone working in an applied field have been and still are said not to be true scientists. Their fields are said to be practical arts where there are no universal principles. "Common sense" rules of thumb are said to be their guides. Only theoreticians are true scientists. This issue derives from the dispute over theory versus history.[30]

Science, whether basic or applied, can be referred to and described as theory—applied sciences do have their own theoretical principles. History requires and uses theory to identify the nature and causes of a specific case, such as sedimentary rock embedded at a forty-five degree angle several thousand feet up on a mountainside or the nature and cause of a specific war. What is called natural history uses theory from the physical sciences, and human history uses theory from the human sciences, to explain the nature and causes of the cases each study. Psychotherapists and medical doctors use the theory of basic sciences to identify and treat their patients' problems, which to the therapist and doctor are historical events.

Economist Ludwig von Mises clarified the theory-history distinction by rejecting the Hegelian-Marxist view that there are laws of history, as well as the historicist view that theory is time- and culture-bound, able only to generate statistical probabilities. Rather, Mises recognized that there is a valid science of individual historical events using theories of the more basic sciences to understand and describe those events. This makes history an applied science, though not a concept of method, or "how-to" discipline, in the way that engineering, basket weaving, or even economics are.[31]

---

[30] This also derives from a condescension of some academics to declare their superiority over laypersons who, so they assume, are incapable of performing, or would perform poorly, any cognitive processes that true scientists perform.

[31] There are, of course, principles and theories of how to do history, called the philosophy of history. Mises's emphasis on the requirement of a valid theory to guide historians is his philosophy of human history, which epistemologically derives from the notion that thought precedes (or should precede) action, which means that thought, good or bad, society-wide or individual, determines historical events. Advocates of the theory of capitalism, subsequently, will produce different historical interpretations of past events than advocates of the theory of socialism. See Ludwig von Mises, *Theory and History: An Interpretation of*

The two fundamental mental processes of cognition are generalization and application. The former is inductive and gives us theory, basic and applied, that is, the concepts and principles of science—and I might add, in the name of the layperson, the concepts and principles that constitute our personal, everyday knowledge. Application is deductive and uses the previously acquired knowledge that generalization gives us and is the fundamental method of applied science. Application is the mental process used in both history and the how-to disciplines. It is what medical doctors do, what Sherlock Holmes did, and what we all do daily. Generalization for many of us, unfortunately, comes to a near halt when we finish our schooling and cease to learn new concepts.[32]

Application, to be more precise, means we identify "a this as an instance of a that," which means *this* specific case or concrete belongs to or is subsumed by *that* universal concept.[33] We present a cough and runny nose to our doctor and he or she quickly concludes, based on accumulated knowledge and patient history, that we have a cold. Similarly, Holmes saw that Watson was tanned and showed signs of having been wounded in a war; thus, he concluded (deduced that) Watson recently came back from Afghanistan.[34] And a child applies the previously learned concept of balance by shifting weight when learning to ride a bicycle. All three examples are processes of deduction and illustrate how deduction is the predominant method of both applied sciences and everyday life.[35]

Social and Economic Evolution (1957; repr. Ludwig von Mises Institute, 2007), pt. 3, esp. chap. 10. Economics is a "how-to" discipline in that its essence is to identify the principles of how to produce wealth. More on that later.

[32] See my preface to *Applying Principles: Short Essays Based on the Philosophy of Ayn Rand, Economics of Ludwig von Mises, and Psychology of Edith Packer* (Kirkpatrick Books, 2021), 11–15. The two processes, as I say in the preface, "are not the monopoly of scientists, philosophers, or academics in general."

[33] Aristotle identifies application as recognition of "particulars actually falling under the universal." Thus, a student who knows from "preexistent knowledge" that a triangle is equal to two right angles will recognize "in the instance [or concrete] before him" that a three-sided figure inscribed in a semicircle will be a triangle. *Posterior Analytics*, 71a20–23, in McKeon, 110.

[34] Arthur Conan Doyle, *A Study in Scarlet: A Detective Story* (1888; repr. SeaWolf Press, 2018), chap. 2.

[35] See my discussion of generalization and application as two essential mental skills that should be taught throughout the education process, starting as early as a child can understand the skills. Jerry Kirkpatrick, *Montessori, Dewey, and Capitalism: Educational Theory for a Free Market in Education* (Kirkpatrick

THE PROBLEM OF UNIVERSALS—HOW DO WE KNOW CONCRETES?

To properly answer the positivists and dislodge them and their followers from their "dogmatic slumbers" of complacent skepticism, we must analyze and answer the so-called problem of universals.[36] Let me now begin by addressing the first relevant question that is at the root of the problem: How do we know concretes?

To the layperson this question must seem strange. To academics, especially philosophers, the issue for a couple of millennia has been a big deal. Answering the question of how we know concretes will answer the problem of universals and at the same time the more general question, how do we know?

Of course, we know concretes, says the layperson. We open our eyes and look out at the world—to see specific, individual concrete people, animals, plants, and things. Reminiscing about our childhood, we can recall many specific, individual concrete events, ranging from games we played with relatives or neighborhood friends to vacations taken with the family to the years spent in various schools. And historians specialize in reporting the concretes of the past, do they not?

To put the issue in more philosophical terms, we live in a world of particulars, yet with the exception of proper names our knowledge consists of universals. Theories, science, and principles are all expressed in terms of universals. How do we use those universals to know particulars?

The problem arises from the theory called moderate realism—Aristotle's improvement on Plato—that says universals are external to our minds, lodged, so to speak, in the things of reality, "out there," but not in Plato's separate realm.[37] Aristotle said that the form is embedded in

---

Books, 2008), 134–42. See Rand, *Objectivist Epistemology*, 28, 50, where she refers to application as a process of "conceptual identification," the process of subsuming concretes under a known concept.

[36] Immanuel Kant wrote that David Hume's failure to find a necessary connection between cause and effect "interrupted [his] dogmatic slumber," requiring him to seek a new direction in philosophy. "Dogmatic" in his time meant a conviction of absolute or apodictic certainty and eternal truth. Hume upset that conviction. Kant's "solution" has been described as complacent skepticism because he was content not to know true reality, only phenomena. He denied that he was a skeptic. I first heard this description of Kant's philosophy from philosopher Leonard Peikoff. Immanuel Kant, *Prolegomena to Any Future Metaphysics*, ed. Lewis White Beck (Liberal Arts Press, 1950), 8. First published in German in 1783.

[37] Plato's theory is called extreme realism. Moderate realism is sometimes denigrated as "naïve," a term that is more often applied to a theory of perception. "Metaphysical realism" is also used to describe Aristotle's theory.

the matter, form being his word for what in more modern terms we call essence or essential distinguishing characteristic. Matter, as in the stuff that something is made of, is in Aristotle's view what makes the thing a unique, concrete individual entity.

How can we know concretes when nearly all knowledge is universal? The glass on my desk, so the theory goes, contains "glassness," the universal essence that we grasp and understand as a thing we drink liquids out of. The soda-lime silicate that the container is made of is its matter, and this includes its color and the number of ounces of liquid it can hold. The matter of the glass, we could say, is that the glass on my desk is made of soda, lime, and silica, its color is blue, and it holds twelve ounces. Does that mean we have understood the concreteness of this glass?

No, say the critics of Aristotle's common-sense realism, because we are using universals in every word ("soda," "lime," "silica," "blue," "twelve ounces") to describe this individual concrete. We fail to get to "the thing in itself" (the thing as it really is, its identity), a problem that led John Locke to conclude that we cannot know the substance, substratum or "pure essence" of a thing that holds its attributes together. Further problems led David Hume to fail to find a "necessary connection" between cause and effect, heading us down the road to Immanuel Kant who said we can never know true, noumenal reality (where the "thing in itself" is presumed to reside). We can only know appearances or phenomena, according to Kant.

And that is where we are today—reality is unknowable, concepts are arbitrary, and nothing is universal—with serious consequences in ethics and politics: values are subjective, because they are not "out there," either, and there is no objectively valid, justifiable political or economic system, such as capitalism. How do we decide who is right if all knowledge is subjective and therefore arbitrary? At best, the positivists say, we can only achieve "intersubjective reliability," never universally true knowledge. Thomas Hobbes, because of his subjectivism in epistemology, said we need a strong "public sword" to keep the peace among warring groups, each of whom seeks to rule using its own arbitrary values. In other words, dictatorship follows from philosophy's failure to solve the problem of universals. Positivists, however, are not disturbed by Hobbes's argument because they believe intersubjective reliability—essentially

---

"Intrinsicism" can refer to both Plato and Aristotle, as they both believe that universals are *intrinsic* to external reality, independent of our method of cognition.

a vote or majority consensus of those working in a particular area—is "good enough," presumably to prevent the establishment of the strong sword to settle disputes.[38]

Let me go back to our layperson who looks out at the world and sees a myriad of concrete things and their attributes. Although such a layperson may still think in Aristotle's realist tradition by saying, "I just look and see glassness in the glass," the conclusion is not correct, because there is in fact no essence of glassness in the thing out there. The layperson's mind is doing more than he or she thinks. And what Aristotle calls abstraction—mentally separating the form from its matter—is a more complicated process than he knew.

Concept formation, or conceptualization, is the human being's method of turning percepts into universal concepts. It is true that essences and values are not out there in a metaphysical sense, but that does not mean they are epistemologically arbitrary. How does this process work?

To this issue, we now turn in chapter 2 with an exposition of Ayn Rand's theory of universals, which she calls a theory of concepts or concept formation.[39]

[38] Practically speaking, this means many citations in peer reviewed academic journals. Peer review, however, is not the criterion of truth. On peer review in science, see Kirkpatrick, *Applying Principles*, 123–32, esp. 130–32 for my criticism of the process from the standpoint of ethics and epistemology.

[39] Academic philosophers treat the problem of universals as strictly a metaphysical question, asking, where are those essences out there in reality, external to consciousness? They fail to find them and, as a result, conclude that knowledge is subjective. Ayn Rand, as we shall see, treats the issue as an epistemological process of forming universals that identify the concretes of reality, both external to our minds and internal.

2

Answering the Problem of Universals

> The process of conceptualization consists of observing the differences and similarities of the existents *within the field of one's awareness* (and organizing them into concepts accordingly).
>
> —Ayn Rand[1]

> The soul is so constituted as to be capable of this process.
>
> —Aristotle[2]

AYN RAND'S THEORY OF CONCEPTS is her most significant contribution to philosophy.

It is more important than her ethics and more important than her political philosophy. Rand's solution to the problem of universals is what makes it possible to validate her ethical theory of rational egoism and her social theory of laissez-faire capitalism, because the foundation of both is her epistemology. To resolve the issue of knowing concretes without contradiction and answer the so-called problem of universals, Rand shifts the problem away from metaphysics to epistemology. She then elaborates how we come to understand abstractions from abstractions and concepts

---

[1] Ayn Rand, *Objectivist Epistemology*, 42. Rand's italics, to emphasize that concepts are formed by individual human beings based on what each attends to or observes.

[2] Aristotle, *Posterior Analytics*, 100a13–14, in McKeon, *Basic Works*, 185.

of consciousness, both of which are essential to developing sound theories of psychology, ethics, politics, and economics, that is, sound theories of the human sciences.[3]

THE INFANT'S EPISTEMOLOGY

Awareness of concretes begins early, perhaps in the womb as the fetus becomes aware of the rudiments and rhythms of the mother's language.[4]

All knowledge as we normally think of it begins at birth with sensation and, in a few months, with the beginnings of perception, by observing concrete particulars related to our experiences of pleasure and pain.[5] Initially, the infant moves its eyes back and forth, carefully observing its mother's face. It cries when hungry and seeks a breast for a feast of milk. When sated, the infant may burp, and later, when noticing digestive sensations in its intestines, it expresses discomfort. After defecation, it may give us a smile.

Though the infant is not thinking in concepts or universals, it is aware through the sensations of sight, taste, and touch of quite a few concrete particulars: mother, mom's breast and breastmilk, my gas and feces. These words are concepts and therefore universal, but to the infant the experiences are unique and individual to its own life and immediate environment.

After two or three months, many sensations become put together automatically by our brains and integrated into percepts, enabling the infant to become aware of (specific) entities: ball, cat, mama, and

---

[3] Additional summaries of Rand's theory can be found here: Leonard Peikoff, *Objectivism: The Philosophy of Ayn Rand* (Dutton, 1991), chap. 3; Jerry Kirkpatrick, *In Defense of Advertising: Arguments from Reason, Ethical Egoism, and Laissez-Faire Capitalism* (1994; paperback ed. TLJ Books, 2007), 147–52; and Kirkpatrick, *Montessori, Dewey, and Capitalism*, 82–86. See also my exposition of Rand's theory and my application of it to the applied science of marketing, "Objectivist Epistemology as the Foundation of Marketing Theory," in Edward W. Younkins, ed., *Perspectives on Ayn Rand's Contributions to Economic and Business Thought* (Rowman & Littlefield, 2018), 223–38. Originally a conference presentation published in C. Whan Park and Daniel C. Smith, eds., *Marketing Theory and Applications*, vol. 5 (American Marketing Association, 1994), 118–25.

[4] Gail Gross, "How Your Baby Learns Language in the Womb," *Psychology Today*, September 28, 2022, psychologytoday.com.

[5] Cf. Kim, Hojin I. and Scott P. Johnson, "Infant Perception." *Encyclopedia Britannica*, May 28, 2014, britannica.com.

even "my fist." Through our senses we perceive concretes, including the similarity of certain concretes. Only later, at age one or so, do we begin to form concepts.

Our world in the beginning of life is concrete and particular, which means perceptual reality, not the knowledge we acquired in graduate school, is the starting point of epistemology. When we begin to talk, we soon learn to form concepts, which become our universals, though we continue to live our entire lives in a world of concretes. The infant likely first says the word "mama," which initially means its own mother. Eventually, it becomes aware that "mama" or "mother" can be a universal that applies to all women who have children. Similarly, it will soon learn "ball" and "kitty."

Before moving on to what we might call the toddler's and young child's epistemology of learning concepts for perceived concretes, let me quickly dispense with the alleged deceptiveness of the senses. When we perceive a bent stick in water, we are experiencing an accurate visual perception. When we put our hands in the water to run them up and down the stick, using our sense of touch, we realize it is not bent. And when we remember our physics of light, we understand that it is the effect of light in water that makes the stick look bent. Our senses do not deceive us. If the stick, using our eyes, looked straight, that would be a deception.

Further, we must look at Ayn Rand's form-object distinction. Animals have many different *forms* of perception while their *object* is still reality, that is, the world "out there." Flies see the world as a mosaic and mostly motion, and many birds with eyes in the sides of their heads view the world differently than we do, called monocular vision. Our eyes are pointed straight ahead, though some of us may be color blind and a few may have an extra cone enabling them to discriminate millions more shades of color than the rest of us. All these different forms of perception are still perceiving a part of reality, giving each of us varying amounts and qualities of information.[6]

[6] Rand rejects the primary-secondary quality dichotomy, because color caused by wavelengths of light does not mean that color is secondary or subjective while wavelengths of light are primary and objective. Color is our *form* of perception. The *object* causes the perception when wavelengths of light impinge on our retinas. We perceive the alleged primary qualities of length and spatial extension, Rand points out, in the same way. Note the analogy of concept formation to perception. A major difference between the two processes is that we integrate, through

The ancient parable of the blind men and the elephant is relevant here. Each blind man touches a different part of the elephant and draws incorrect conclusions about what he is touching. One says the body is a wall, another says the tusk is a spear, and others say the leg is a tree, the ear a fan, and the tail a rope. Absent the sense of sight, the men are handicapped, but further touching could have resolved their errors. We each perceive a portion of reality and our judgments, as in the case of the blind men, using previously formed concepts, may or may not be correct. None of us is omniscient and infallible, nor will any of us ever achieve such acclaim. Much testing and trying is required before drawing conclusions, touching all sides of the elephant, for example, before saying what we think it is.[7]

CONCEPT FORMATION OF PERCEIVED CONCRETES

The mental process of forming concepts, according to Ayn Rand, proceeds as follows. We perceive many differences and similarities among existents, "existent" meaning anything that exists—things or entities, attributes, actions, or relationships of entities in the world. When we focus on two or more existents that are similar to one another, but also different from other existents, we have isolated something we want to identify and classify with a concept and word. It is inherent in the nature of human consciousness, says Rand, to be able to view entities and other existents as units, meaning as individual members of a group of similar members. Other animals cannot do this, at least not in the same way or to the same extent as humans can.

Measurement Omission

To form the concept, we focus on the fundamental characteristic or characteristics using Rand's rule of fundamentality—the one or ones that explain and cause all or most of the other characteristics. For example, an animal possessing the capacity to reason is fundamental as the essence of human being, whereas opposable thumb or ability to build skyscrapers as the differentia is not. In the process of

volitional effort, the data of our percepts into concepts by means of the native inborn method called conceptualization. See Rand *Objectivist Epistemology*, 179–82 and Leonard Peikoff, *Objectivism*, 39–54.

[7] The blind men with their elephant were mistaken because they did not exert further effort to touch more of the object before drawing a conclusion. A good lesson or metaphor for any kind of learning.

identifying the fundamental, we omit its measurements. Measurement omission is the mental process of abstraction that enables us to identify the essence of a concept. The concept of essence refers to the same characteristic of several similar existents out there in reality, but in our minds it is absent the characteristic's individual measurements. Similarity means several observed existents sharing the same characteristic with their differing measurements ignored. For "human being," we omit the content and intensity of each individual's reasoning capacity.

These measurements must exist in some quantity, as all aspects of the universe are quantitative, and be commensurable—shape, for example, for tables, and hue for colors—but the omitted measurements are neither disposed of nor forgotten. Conceptualization is implicit or approximate measurement, meaning we do not need to know how precisely to measure the referents in order to form a concept. We formed the concepts of color long before we knew how to measure the wavelengths of light that cause the colors.[8] But we do need to bring the measurements back in to apply the concept to particular cases, especially in the applied sciences.[9]

Note that the word "explain" as used here is epistemological and "cause" is metaphysical. Epistemological explanations occur in a mental process, sometimes externalized as speech or even in published articles and books. Metaphysical events and causes occur outside or independent of that mental process.

Using Rand's example of forming the concept "table," we focus on the commensurable characteristic of the shape of tables, chairs, and beds and isolate tables from chairs and beds by their particular shape and function, namely as objects that have flat surfaces and are designed to hold smaller objects, but not to sit or sleep on. Individual, concrete tables differ according to their measurements, but the precise measurements of their height, length, width, color, oval versus rectangular, etc., are omitted to make the concept "table."

---

[8] Thus, Rand's definition of concept: "a mental integration of two or more units possessing the same distinguishing characteristic(s), with their particular measurements omitted." Italics from Rand's text omitted. Rand, *Objectivist Epistemology*, 13.

[9] "Individual differences," a term used by psychologists, refers to the measurements that are brought back in to identify and describe particular "concrete" human beings.

The omitted measurements, again, are still there, in reality in the identity of the entity and in the concept formation process operating in our minds, but they are put in the background to form the essence of the concept. Identifying that tables are flat and hold other objects is all we need to know to distinguish large and small tables from chairs and beds and to give us the essence or essential distinguishing characteristic of tableness. Particular height, width, weight, color, shape of top, etc., are left out of the formation process, but not of the concept itself.

The File Folder Metaphor

Rand's metaphor of concepts as file folders emphasizes that every observation we have had of a concrete is retained in the concept (the folder) but in the background.[10] The word "table" in the English language is assigned to the concept—written, say, on the tab of the folder—and its definition, genus and differentia, would be something like "furniture with a flat, level surface and support designed to hold smaller objects," written on the outside of the folder to indicate the essence and summary of the folder's content.[11]

The "folders" we have in our minds vary from person to person in content and "thickness," because of our varying experience with and knowledge of tables. The carpenter who specializes in making tables for a living has a "fatter" folder than the rest of us. And children have a thinner folder of "human" than adults do, which means concepts are formed within one's own state of knowledge, as well as within the state of humankind's knowledge, and can be adjusted as knowledge expands. The definition is a condensation of the concept's content, not something tautological or equivalent to or interchangeable with the concept, as the nominalists and positivists insist.[12] For Rand the definition summarizes and condenses, which means all accumulated knowledge in that folder is retained in our subconscious minds when we use the concept.[13]

---

[10] The word "retain" must be used advisedly, as our memories may be clear and sharp or fuzzy and dim.

[11] Cf. Rand's definition of table, *Objectivist Epistemology*, 11.

[12] Rand, 96–97.

[13] And how well we retain and use that accumulated knowledge depends on how well we have maintained and ordered that content—clear and sharp or fuzzy and dim.

Axiomatic Concepts

All the varying aspects of each entity in the world, including the varying measurements of its attributes, and including its location and time of existence, are what give each entity its unique individuality or concreteness. The entity—its identity—*is its attributes*, all of them. Identity is a corollary of Rand's primary concept of existence. She calls such primary concepts *axiomatic*, meaning they are implicit in all knowledge and cannot be denied without assuming them in the process of denial. When stated as a proposition, the axiomatic concept of existence says, "existence exists," and its corollary says, "existence is identity." This is an updated restatement of Aristotle's primacy-of-existence premise.[14]

In discussions of induction, the uniformity-of-nature principle is an expression of Aristotle's law of identity and, as such, is unnecessary. If anything, it complicates discussion. The proper understanding of the law of identity also eliminates the need to discuss "complete enumeration." Concepts are not formed by examining every referent. We look at the entity's attributes and, out of cognitive necessity, select the one or more that explains and causes all or most of the rest.[15]

Rand's other primary, axiomatic concept is consciousness, and it says, when stated as a proposition, "consciousness is conscious"—of existence, which means "consciousness is identification" of reality.[16] Descartes's primacy-of-consciousness premise, the prior certainty of consciousness, that has plagued philosophy since his time, is rejected in this simple statement. Rand's response to Descartes's *cogito*, "I think, therefore I am," is her primacy-of-existence proposition, "I am, therefore I'll think."[17]

Primacy of existence means existence exists and consciousness is conscious—of existence. "Prior certainty," to use Descartes's words, is of existence, not consciousness. Or, rather, existence exists, whether or not we are here to perceive it. "If nothing exists," says Rand, "there can be no consciousness: a consciousness with nothing to be conscious of is a contradiction in terms." So also, "A consciousness conscious of

---

[14] Rand, *Objectivist Epistemology*, chap. 6. Ayn Rand, *Atlas Shrugged* (Random House, 1957), 1016.

[15] Cf. McCaskey, "Induction in the Socratic Tradition," 182–83.

[16] Rand, *Objectivist Epistemology*, chap. 6. Rand, *Atlas Shrugged*, 1016.

[17] Rand, *Atlas Shrugged*, 1058.

nothing but itself is a contradiction in terms."[18] Consciousness evolved biologically to be our means of awareness of reality (existence), which includes consciousness itself, and to guide our choices and actions.

Knowing the Concrete

How do we know the entity as a concrete? As said before, initially through perception, which includes the awareness of similarity. How do we know to omit those measurements to come up with the universal concept? As Aristotle said, "The soul is so constituted as to be capable of this process." It is in the nature—identity—of a human consciousness to do this, often without even thinking about it. Children do not think about it, nor do adults when focused on a concrete object; such objects are often defined ostensively by pointing. It is, incidentally, in this sense that advocates of moderate realism tend to say "we just see it" out there in reality. We do, sort of, but only in the sense of omitting measurements without a great deal of effort. When we move from directly perceived concrete objects to abstraction from abstractions, we will be exercising significant volitional effort, to be discussed below.

Aristotle the empiricist states the process correctly, namely that all human knowledge begins with sense-perception. Such perceptions, in his words, are initially retained as memories, then several memories form an experience, and finally, a universal is comprehended in the soul. This "comprehension" occurs through *nous*, our reasoning mind. Unfortunately, Aristotle's *nous* in this context is often translated as *intuition*, implying a mystical force or activity.[19] For Aristotle the mental process is the grasp of the universal in the individual, the form in the matter. *Nous* in this sense most likely means a type of intellectual power that is capable of moving from awareness of the individual to the universal. Hence, Aristotle's line about the mind being "so constituted."[20]

---

[18] Rand, *Atlas Shrugged*, 1015–16.

[19] "So out of sense-perception comes to be what we call memory, and out of frequently repeated memories of the same thing develops experience; for a number of memories constitute a single experience." Aristotle, *Posterior Analytics*, 100a3–9, in McKeon, *Basic Works*, 185. In the last paragraph of this chapter, 100b5–15, Jonathan Barnes, ed., *The Complete Works of Aristotle*, revised Oxford translation (Princeton University Press, 1984), 1:166, translates *nous* as comprehension.

[20] Scholars debate the meaning and translation of Aristotle's simile of a rout in battle, one sentence before the noteworthy one: "It is like a rout in battle stopped by first one man making a stand and then another, until the original formation has been restored. The soul is so constituted as to be capable of this

Aristotle did not know the problems that would arise from his moderate realism or about the notion of measurement omission, but he did give us insight into the process. Rand now gives us the more fully correct identification. Updating Aristotle's sentence, we might say: the mind is so constituted as to be capable of omitting measurements in order to identify epistemological essences.

Measurement omission is what makes our concepts universal. By omitting the differences among the similar concretes, especially their quantifiable differences, we now have something in our minds, the essential distinguishing characteristic, that identifies the nature of any existent of that kind, past, present, and future, which means our concept is open-ended. *All* tables are flat with supports to hold smaller objects. Our concept is not corresponding to or matching a substratum or intrinsic essence "out there," but *identifying* what all tables in a *qualitatively* measurements-omitted sense have in common.

The Theory of Truth

For Rand, consciousness is identification, which means that her theory of truth is an identification theory, not, strictly speaking, one of correspondence.[21] To identify means to recognize similarity and to subsume or classify the similar entities under one idea.[22] The idea becomes integrated into a mental entity called a concept and is further

process." Aristotle, *Posterior Analytics*, 100a11–14, in McKeon, *Basic Works*, 185. It seems, though, that Aristotle is simply comparing the "rout" to the operation of our minds beginning with a perception, then moving to several memories that become an experience to form the universal—like one man returning to the battle line, then another, then all to achieve a "position of strength," as others have translated the last few words. Cf. James Lesher, "A Note on the Simile of the Rout in the *Posterior Analytics* ii 19," *Ancient Philosophy* 31, no. 1 (2011): 121–25, for his discussion of "rout" as used by Aristotle in other works.

[21] "Truth is the product of the recognition (i.e., identification) of the facts of reality." Rand, *Objectivist Epistemology*, 48. Rand does call her theory a correspondence theory of truth, but only in the sense that correspondence means agreement with reality, resulting from a specific method, rather than a reflection or photograph of the essence "out there." It is not a "coherence" theory, favorite of the German and British idealists.

[22] "To subsume (several things) under one idea or concept (1610)," *Oxford English Dictionary*, s.v. "identify (v.), Etymology," June 2024, https://doi.org/10.1093/OED/1204309056. In other words, as used above, to identify means to classify. This usage differs from the times we recognize or identify a particular person as John, a place as Los Angeles, or a thing as *The Fountainhead*, using a proper noun.

identified with a word.[23] The epistemological (in here) concept *means* its metaphysical (out there) referents—the concept table means all of the large and small, round and oval tables we have observed, or will ever observe or not observe. Our ever-expanding accumulation of concepts—our increasing knowledge—enables us to know and describe our observed concretes in their extensive variety, their attributes and actions and their myriad differences according to their varying measurements.

Our minds are not passive "mirrors of nature," words critics use to attack Aristotle's moderate realism.[24] They are active processors of it—in an analogous way that our brains process sense data to form percepts. Rand's recognition of the mind's active nature and her incorporation of it into her theory provides a major improvement on and defense of Aristotle's epistemology. Her theory has been called *epistemological* realism in the acquisition of knowledge.[25]

Assuming our concepts accurately describe the concretes they identify, not contradicting them, and are not disconnected, detached, or floating, we have objective knowledge.

And that objective knowledge can be basic science or applied science, theory or history, general knowledge or our own personal everyday knowledge. In all cases, it is knowledge of the concretes of reality.

---

[23] The use of different words to identify similar, but subtly different concepts, indicates the range of options in concept formation, most evident in the challenge of translation from one language to another. Tribal languages, interestingly, often have fewer concepts of color, four in one language with the concepts denoting a wider range of frequencies than Newton's seven. Another tribe has only two, likely related to "warm" and "cool" temperatures as their meanings. Still another tribe has a word for every plant and animal in its habitat, but no general concept for plant or animal. On so-called borderline cases, see Rand, *Objectivist Epistemology*, 72–74. Concepts are formed based on cognitive need.

[24] Other words used to describe the moderate-realist view of consciousness are "reflective" and "diaphanous." Our minds are said to reflect nature because consciousness is transparent and, consequently, has no identity. See Norman Kemp Smith, in *A Commentary to Kant's "Critique of Pure Reason"* (Macmillan, 1918), xl–xliii. For Kant, consciousness is active and has a nature—at the expense of awareness of reality.

[25] David Kelley, "Concepts and Propositions," 3, *The Atlas Society*, June 6, 2025, atlassociety.org. Rand denied that her theory was a form of conceptualism, the theory of Scholastic Peter Abelard and British empiricist John Locke, both of whom failed to explain how concepts in our minds relate to and identify their referents in the perceived concretes of reality. Ayn Rand, however, seems to have succeeded in doing just that.

ABSTRACTION FROM ABSTRACTIONS

The significant advantage of the human form of cognition, as opposed to that of higher animals who exhibit a high level of perception (and perhaps a modicum of choice), is that conceptualization identifies and integrates existents into a single unit—the concept—and thereby expands our knowledge beyond the directly perceived concretes, which the higher animals are limited to.

Unit Economy

Rand calls this process the principle of unit economy. Just as mathematical measurement reduces lengths to perceived units, such as foot, yard, and mile, and enables us to project and comprehend great distances, such as the length of travel from the earth to the moon, non-mathematical concepts do the same. We cannot perceive the actual distance between the earth and the moon, but we can *conceive* it. Similarly, we cannot perceive all tables past, present, and future, or easily hold in our minds all the ones we have perceived, but we can conceive of and retain their genus and differentia, knowing that when we come upon some human-made object that is flat and holds smaller objects it will be a table.

The concept reduces all these perceived concretes to a single symbol, the word, which when automatized and retained as representing the meaning and referents of the concept, enables us to recall and use the content of the concept in the present. The concept "5," Rand points out, does not require us to perceive five sticks or fingers every time we use the word, which we had to do when young to learn the concept's meaning.[26] All well-defined concepts, which means well-organized knowledge stored in our subconscious minds, work the same way. They give instantaneous understanding that we can recall and use at will. Poorly defined concepts, however, and knowledge in disarray, including false and invalid concepts, introduce chaos into the operations

[26] Rand, *Objectivist Epistemology*, 63. Mathematics is in fact a highly deductive science, once we empirically induce its *basic* concepts and operations. That is, once we know the concepts five, two, plus, sum, and seven, we can deduce the sum of five and two as seven. My emphasis is on the concept "basic," because once we learn the basics, we deduce the higher numbers and processes. We do not have to observe fifty sticks to comprehend the concept "fifty" or stacks of ten each to understand addition. Mathematics is empirical, not "a priori," prior to experience.

of our minds. The principle of unit economy explains the necessity of good mental housekeeping when moving from perceived concretes to abstraction from abstractions.

Unit economy is the tool of abstraction from abstractions that enables us to form, comprehend, and apply the broadest and narrowest of philosophic and scientific terms, often involving lengthy chains of abstraction. If these concepts are not solidly tied to perceived concretes from which they arise, by tracing their connecting links, they will be floating and one's understanding of them will be fuzzy and approximate, or even false. This is where volitional effort is required to form and maintain the accuracy and understanding of our knowledge.

Wider and Narrower Abstractions

The process of abstraction from abstractions moves in two directions, to wider and wider concepts and to narrower subdivisions of previously formed concepts.

When we form the wider concept of furniture, a first-level abstraction from abstractions, we integrate the concepts of table, chair, and bed, our concepts of perceived concretes, by omitting their differences in shape and identifying their common purpose as large objects to support smaller objects or the human body as the essential distinguishing characteristic of furniture. Similarly, we can go wider and integrate furniture with furnaces, air conditioners, and kitchen appliances to form the concept of household goods. And we can go still wider with our abstraction from abstractions and integrate household goods, bridges, automobiles, medicines, jewelry, and skyscrapers into a concept of human-made objects. The process is to omit the distinguishing characteristic of each individual item in the initial abstractions we are attempting to integrate into a wider concept and focus on what they have in common to form the broader concept. All characteristics of the former are nevertheless retained in the new concept.[27]

The concepts of tables, chairs, and beds are now mental entities that are metaphorically treated as concretes or referents of the concept furniture. Similarly, furniture, furnace, air conditioner, etc., become mental entities and referents of the concept household goods. This chain of abstraction is what must be clearly understood in the formation of our concepts, lest it all become confusing or misidentified. The

[27] Rand, *Objectivist Epistemology*, 21–23.

notion of abstraction from abstractions means our knowledge exists in a hierarchical context, which means that some concepts presuppose others in the chain of abstractions.

We can narrow or subdivide concepts, which, like the wider abstractions, move further away from the directly perceived concrete. The concept table, for example, can be subdivided into kitchen, office, and living room tables, with each new concept retaining the distinguishing characteristic of a flat object to hold smaller objects, but narrowing it by the nature of the respective table shapes and functions. We can narrow the concept of human being to many different subclassifications, such as subdivision by age: child, adolescent, adult; or by occupation: engineer, doctor, truck driver. The perceived concrete remains the human being—or living thing that exhibits the capacity to reason—and is presupposed by the formation of these narrowed concepts.[28]

The challenge occurs when forming higher level, complex concepts, such as concepts of grammar, concepts of consciousness, and axiomatic concepts, which Rand discusses in detail.[29] Composite concepts, as Rand calls some broad abstractions, require the integration of concepts of consciousness with existential concepts. Marriage, a relationship between two people, usually a man and woman, for example, presupposes the mental processes of love and commitment, plus the concepts of contractual agreement, morality, and law. Property, a relationship to an object or land, presupposes the concepts of the means of acquisition (purchase, rental, or gift), morality, law, and rights. Justice, a relation of one person to another (or to oneself), presupposes the concepts of judgment (of oneself or others) based on a criterion rooted in ethics, especially the concepts of right and wrong, or good or bad. Marriage, property, and justice in each respective case requires further tracing of the chain of abstractions to its perceived concretes, which with these concepts mean one or more human beings and their actions in various relationships.[30]

Proof, scientific or otherwise, depends on our prior knowledge and that prior knowledge is only as good as the concepts and principles (which are combinations of concepts) we use. And that knowledge must be well-organized in order to retrieve it for understanding and proof.

---

[28] Rand, 23–26.

[29] Rand, 37–38, 181–83, chap. 4, 6.

[30] Rand, 36–37, 51–52. If we identify precisely what a broad abstraction *presupposes*, we are following or tracing the chain of abstractions.

If we can identify precisely the concepts that a wider or narrower abstraction *presupposes*, we will know that we are following or tracing the correct chain of abstractions. All connections in these chains, wider and narrower, must be traced to the perceived concretes on which they are ultimately based. This can be described as a modernized Aristotelian empiricism.

The Well-Organized Filing System

To maintain order in our minds requires good—meaning concise and true—definitions. Definitions summarize and condense the content of the file folders, our concepts, and denote the concept's referents by naming the essence, through a genus and differentia. The genus connects each concept to the broader category from which it was differentiated and the differentia identifies it.[31] Each definition uses numerous previously identified and defined concepts. Conceptualization, as a result, as Rand points out, is a complicated and elaborate system of classification, of "mental filing and cross-filing."[32]

Our concepts, whether of a perceived concrete or a broad abstraction, are, or should be, contextually defined and hierarchically related to those they rest on or presuppose.

CONCEPTS OF CONSCIOUSNESS

As a natural attribute of the entity homo sapiens, the human consciousness is not just a warehouse of accumulated knowledge, but more importantly an active processor of the data of external and internal reality to acquire, store, and apply that knowledge. Observation is the fundamental method of scientists and laypersons alike in the identification of essential distinguishing characteristics to form

---

[31] In addition to genus and differentia, and stating the concept's essence, the rules of a good definition caution that the definition must not be too broad or narrow, circular, vague, obscure, metaphorical, or negative, unless the concept is essentially negative. See David Kelley, *The Art of Reasoning* (W. W. Norton, 1988), 32–43.

[32] Rand, *Objectivist Epistemology*, 69. I suggested in chapter 1 of the present work, p. 19, that scientific experimentation likely derived as an analogy from our form of conceptual thinking. I would suggest here that the intricate filing and cross-filing of our concepts provided the analogy to the development of computers. Our minds are not like computers; computers are like our minds, albeit limited in what they can mimic. Cf. John Searle, *Minds, Brains, and Science* (Harvard University Press, 1984), 44. Throughout history, Searle points out, the latest technology becomes the metaphor of the brain: catapult for some Greeks, a mill for Leibniz, and a telephone switchboard for the pre-computer twentieth century.

concepts. More specifically, extrospection is the method used for external reality, introspection for the internal.[33] Thus, consciousness exhibits a number of processes that need to be identified. Here is a brief summary of how basic concepts of consciousness are formed. Subsequently, I will expand on what I mentioned earlier as concepts of method, as these become relevant in our discussions of the human sciences, especially psychology and economics.[34]

Content and action of mental processes are the two aspects we focus on to form concepts of consciousness. Both are quantitative. Mental content comes in different amounts of knowledge and also varies in length of the conceptual chain required to form the content. Mental actions have varying degrees of intensity, based on such dimensions as scope, clarity, motivation, and mental energy, all measured approximately since there is no precise method of measuring any aspect of consciousness. The way we form concepts of mental processes is to omit the specific content and degree of intensity, then identify what the process performs for us. Thought, as Rand states, is a "purposefully directed process of cognition," and emotion is an automatic, psychosomatic response to an object, person, or event according to a previously formed value judgment.[35] Other examples that Rand defines are perception, evaluation, reminiscence, imagination, and love.

The human consciousness involves and uses many mental processes that need to be identified without contradiction. By mental process, I mean an identifiable action of consciousness, a power or ability we possess by the nature of being a human being, to respond to existence and ultimately to guide our choices and actions. The "perceived concrete" of a concept of consciousness, which we become aware of through introspection, is a single instance of the mental process—a single thought, emotion, memory, and so on.[36]

Concepts of method—specific principles of guidance, as I called them earlier—are omnipresent. They define a goal and the steps or principles to guide us in how to achieve that goal. These concepts of

---

[33] Cf. Kirkpatrick, *Independent Judgment*, 140–41

[34] Rand, *Objectivist Epistemology*, chap. 4.

[35] Rand, 32. The definition of "emotion" is Edith Packer's slightly expanded version of Rand's. Edith Packer, *Lectures on Psychology: A Guide to Understanding Your Emotions* (2012; repr. TJS Books, 2018), 223. Cf. Kirkpatrick, *Independent Judgment*, 63.

[36] Rand, *Objectivist Epistemology*, 31.

method include not just epistemology, logic, and engineering, but also marketing, basket weaving, and the making of a toy boat. The concepts are formed by identifying the goal and course of action, omitting the measurements of both. Epistemology, says Rand, focuses on how we acquire and validate knowledge, while logic is the "art of non-contradictory identification."[37]

Not surprisingly, Rand states that all applied sciences focus on the discovery of methods, which means they are concepts of method. Once the universal concepts of an applied science are discovered and learned, the practitioner then applies the universals to a particular case, say, to build a bridge, sell a product, or weave a basket. The mental process at this last stage is application.

Thus, the solution to the so-called problem of universals is that essences, the universals, are not "out there" metaphysically in external reality. They are epistemological, meaning in our minds, created by us by identifying similarities among concretes and omitting the referents' ranges of measurable differences. If identified correctly, by using Aristotle's law of non-contradiction as guide, the universals—our concepts—will be *epistemologically* objective.

THE METAPHYSICAL VERSUS THE EPISTEMOLOGICAL

Ayn Rand's theory of concepts, based on Aristotle's fundamental premise of the primacy of existence, makes an important distinction between the metaphysical and the epistemological.[38] The subject is technical and polemical, but needs to be emphasized and clarified, especially to take us out of the weeds and quagmires of academic philosophy that have troubled epistemology and psychology for several centuries.

Objects of Awareness versus Means of Awareness

Metaphysics, in contrast to the special sciences, studies all of existence—reality, the universe as a whole. This includes consciousness, which is a natural, biological, not supernatural, part of existence that is an attribute of many higher-level animals, especially human beings. Metaphysics studies the fundamental nature of human consciousness,

---

[37] Rand, *Objectivist Epistemology*, 36. "Art," when used to describe a "how-to" discipline, means applied science that provides principles of guidance to the achievement of a specific goal. It does not mean *imprecise* or *unscientific*. Or "just" *common sense*.

[38] Rand, *Objectivist Epistemology*, chap. 1, 2.

while epistemology studies our methods of knowing, that is, the mental processes by which we know, or attempt to know, existence. Rand identifies two primary, self-evident concepts: existence and consciousness, both self-evident in the Aristotelian sense that they are implicitly known in every act of awareness and thus cannot be denied without having to be accepted in the process of denial. Hence, such denials are self-contradictions. These are the two primaries we saw earlier that she calls axiomatic concepts. Identity, which is implicit in and corollary of the fact of existence, is the third axiomatic concept.[39]

When Rand uses the words "metaphysics" and "epistemology" as describing adjectives, such as metaphysical referent and epistemological essence, she narrows the usages to draw attention to how her theory of concepts differs from the moderate realism of Aristotle and from the nominalism, or arbitrary subjectivism, of today.[40] The metaphysical, in its older, traditional sense, is everything that is outside of our minds—termed objective reality—and everything in or related to consciousness is termed subjective. It would be better, I submit, to use the term "external" as modifier of "outer" reality since consciousness is our internal, *metaphysical* reality when it becomes the object of study to formulate theories of epistemology and psychology.

In Ayn Rand's theory, there are metaphysical *and* epistemological components of consciousness. Epistemological, or psychological, a word that Rand sometimes substitutes for "epistemological," is everything our minds do—the mental processes they perform to create mental products, that is, accurate concepts and principles used to acquire knowledge of both the external world and the internal world of consciousness, as well as to guide our choices and actions. Rand here is emphasizing the difference between perceiver, the method of using our minds, and perceived, the facts of reality. When the processes and products of our minds are the objects of study, our consciousness is the perceiver, and the processes and products are the perceived.[41] We are at once both observer and observed. How can we do this and still arrive at valid, objective knowledge?

"Epistemological" as the describing adjective refers to our use of the mental processes and products of our minds, such as conceptualization (the mental process) to form the concept "table" (the mental product),

[39] Rand, chap. 6.

[40] Rand, 21, 35, 52.

[41] Rand, 165–66. On her use of "psychological," see Rand, chap. 4 and p. 256.

or, for that matter, the concept "concept" (another mental product).[42] We are being "epistemological" whether extrospecting or introspecting, which is to say that we are being epistemological when formulating those sciences of consciousness, epistemology and psychology, as well as all the sciences of matter. The objects of study, the *referents* of the concept "table" and of the concept "concept" are *metaphysical*, as are the consciousnesses of each one of us when under study by an epistemologist or psychologist. The *essence* of each concept is epistemological.

The relation between "process" and "product" is that mental processes are actions of consciousness that produce or generate mental products—end results of the processes such as concepts, principles, emotions, values, and bodies of knowledge.[43]

For years, if not centuries, the subject-object distinction in both philosophy and science has tormented all those who work with consciousness. However, it is not a contradiction to say that the processes and products of our minds are the objective, metaphysical reality that both epistemology and psychology study. They are "subjective" only in the sense they are "in the subject," the subject's mind, as opposed to being "out there" in the external world. Consciousness is our internal reality and also is our means of knowing existence—all of existence, including consciousness itself.

The metaphysical is our object of awareness, which includes consciousness, while the epistemological is our means of awareness—of both external and internal reality.

The Double Duty of Studying Consciousness

Epistemologists and psychologists, in effect, perform a "double duty," of using the epistemological methods of our conscious processes to identify the metaphysical nature of those processes. "Double duty" means scientists of the mind must introspect, that is, use their minds to identify their minds, their mental content and processes, as opposed to the scientists of matter who extrospect—in a less complicated way than that of the scientists of the mind—to identify non-mental entities, such as rocks, planets, and molecules.

---

[42] Rand, 230.

[43] The notions of mental process and product are analogous to manufacturing *processes* that create an end *product* or finished good.

It is equivocation to say that psychology is subjective because its object of study is "in the head," or for that matter, to say that essences and values are subjective because they are not "out there" in external reality. It is an unjustified leap to conclude; because of those challenges of identifying the nature and processes of consciousness, that knowledge of consciousness can never be objectively known, meaning that whatever is asserted must be false, arbitrary, or unknowable, and, in particular, distorting of "true" reality.[44]

This is precisely what the Kantians and post-Kantians assert, namely that because our consciousness has an active nature and we must somehow look inward to know consciousness itself, we cannot know, or know with certainty, what is inside or even outside of our minds. Consciousness, they say, distorts awareness of reality.[45] How can they know that? Yes, some people's minds distort reality, such as the ignorant and neurotic, and everyone at times makes mistakes, but there is a solution: follow Aristotle's laws of logic. Mistakes presuppose a reality to know and that there exists a mental process by which those mistakes are corrected. Introspection has been effectively banned from psychology for over a century, but the ban ignores that logic is "*the* introspective science." The so-called argument against introspection presupposes the validity of introspection.[46]

---

[44] The equivocation occurs on both "objective" and "subjective." Objective can mean (1) the *metaphysical* referents of existence, the outer physical world *and* the inner mental world, or (2) the *epistemological* essences we correctly formulate to identify *objective* truth and knowledge. If our concepts and essences—as well as our values—are true, they are epistemologically objective. Only if arbitrary or false would it be correct to call them subjective. Which means we must be careful not to confuse the two meanings of "subjective": (1) in our heads, as one meaning, or (2) epistemologically arbitrary or false, as the other.

[45] The confusion between metaphysical and epistemological is at the root of Heisenberg's so-called uncertainty principle. Just because we do not *know* the speed and direction of a microscopic particle does not mean that the particle has no real speed or direction. See Peikoff, *Objectivism*, 18–19.

[46] Kirkpatrick, *Independent Judgment*, 65. My italics. Introspection is the method used by all idealists, from Plato to Kant and the present, at the expense of reality. They arbitrarily assert the prior certainty of consciousness, thereby invalidating extrospection and reality. Materialists then step in and say, "but we need reality and extrospection." Therefore, introspection is invalid, and consciousness is an illusion. This is the mind-body split codified in philosophy. The solution is to bring mind and body back together using Aristotle's primacy of existence.

The criteria of epistemological objectivity are Aristotle's laws, especially the law of non-contradiction. If I say that the contents of my mind include seeing the object on my desk as a mischievous little bearded shoemaker (a leprechaun), not a glass of water, there is a contradiction between what I am thinking and what the actual fact is.

A valid theory of concepts, or universals, must answer the question, how exactly do the products of our consciousness relate to the facts of both external and internal reality? How do we know that what is "in our heads," epistemologically, correctly identifies what is "out there," metaphysically. And in a more complicated fashion, how do we know that what is "in here," in our own minds, that is, its processes and products, correctly identifies what is . . ."in here"?

Ayn Rand's theory of concepts holds that concepts and essential distinguishing characteristics are in our minds, not "out there" in the thing, but if we correctly identify what is "out there," and "in here," the essences and concepts are objective in the epistemological sense, but not intrinsic or metaphysical as both Plato and Aristotle thought. Historically, philosophers have exerted considerable effort trying to find essences "out there" in the thing, but never found them, often concluding that concepts or universals are subjective "names" only, hence the theory of nominalism.

That the essences are not intrinsically embedded in the things of metaphysical reality does not mean that they do not or cannot have a valid existence elsewhere, epistemologically, in our minds. We are not condemned to skepticism, in particular not to Kant's and the positivists' complacent skepticism.

Indeed, any doctrine of skepticism, ancient or modern, is self-contradictory, because its proponents, in effect, and often explicitly, say, "We know that we know nothing," which is absurd. And the statements of the complacent skeptics—"we don't need complete certainty to live our lives"—are equally absurd. The concepts of probability and uncertainty presuppose certainty in the same way as the concept orphan presupposes parent—that is, we cannot talk about orphans without assuming the existence of parents. Similarly, we cannot talk about a distortion of reality, or an "unknowable" reality, before we have a correct and valid concept of reality. Rand calls this the fallacy of the stolen concept. In all cases, the fallacy ignores the hierarchical structure of abstractions from abstractions.

To elaborate, certain concepts are said to be stolen because they are taken out of the hierarchical order in which they were formed. Users of the fallacy assume a concept they have no right to assume—certainty and reality, for example, both of which they use in their attempts to deny them.[47] In addition, the concept unknow*able*, with emphasis on the suffix, is invalid. There is much we do not know, but that does not mean our minds are *incapable* of knowing reality. When skeptics assert that we "don't need complete certainty," the word "complete" is unnecessary because certainty is the "parent" concept of and is distinguished from such "child" concepts (narrowed abstractions) as possible, probable, and hypothetical *certainty*. "Apodictic" is a word Kant uses (as does Ludwig von Mises) and a standard synonym is "absolute," but those words add nothing to the meaning of certainty. The Greek word *apodeiktos* means "clearly demonstrated," and without clear demonstration, that is, solid evidence, there is no certainty.

To sum up, the metaphysical refers to existence, which includes the processes and products of consciousness. The epistemological refers to consciousness as an active processor, both extrospectively and introspectively, guided by Aristotelian logic, that correctly (or incorrectly) identifies the reality of the "out there" and the "in here."

The Meaning of Logic

Which leads to a final point about logic. The non-contradiction criterion, as I have said several times, means that the "in here" contents of consciousness correctly identify (match or correspond to, provided we do not take these words too literally) the facts of reality, facts that may be either "out there" or "in here."

Ayn Rand's formulation of logic as "non-contradictory identification," I submit, focusing on the "in here" in relation to the "out there," radically challenges the tradition of logic textbooks that have concentrated, in some cases exhaustively, on the deductive logic of syllogism.

[47] Nathaniel Branden, "The Stolen Concept," *The Objectivist Newsletter* 2, no. 1 (1963): 2, 4; Rand, *Objectivist Epistemology*, 59–60. The fallacy is related to, but not the same as, begging the question. Stolen concept assumes what one is attempting to deny or ignores the presupposed concept altogether. Begging the question assumes what one is attempting to prove.

Over the centuries this has given us both Scholastic and modern rationalism that is obsessed with deductive reasoning disconnected from the facts—such as endless debates about angels on the head of a pin, windowless monads, or brains in a vat.[48] The rules and validity of syllogism, a great discovery of Aristotle's, are valuable to know in the construction of valid and true knowledge. And this includes discussions of contradictions in terms and argument consistency.

Excessive attention, however, to an invalid syllogism that says all dogs and cats have four legs and therefore are the same, ignores or downplays a better, sharper attention focused on the referents of each concept—dogs, cats, and other four-legged animals—to identify their respective essential distinguishing characteristics.[49]

In addition to the rationalism of the textbooks, I also question the excessive use by logicians of the word "proposition," because I think it feeds into their rationalistic thinking. Aristotle's word in the Greek is *protasis* and is usually translated as proposition, though the Liddell and Scott *Greek-English Lexicon* adds that the word in particular means a premise in a syllogism. Today's usage makes it a true or false statement.

My complaint is not that traditional logic textbooks are not worth reading—I have read three—I think the excessive use of proposition overlooks the significance of universals, especially thinking in universal principles, and their non-contradictory identification of the facts as Ayn Rand essentializes logic. The proposition and sentence "water with no impurities at sea level boils at 212° Fahrenheit" is not just a proposition or statement; it is a universal principle, with the word "universal" being redundant.

---

[48] Scholastics apparently did not debate angels on the head of a pin, though Thomas Aquinas discussed the spiritual nature of angels and questioned whether they could occupy space. The expression is later satirization of how Scholastics did in fact debate—endless Euclidean-style deductions pro and con from accepted authorities. Windowless, sometimes called "naked," monads are Leibniz's fundamental substances of the universe. And "brains in a vat" is the twentieth century's absurd, arbitrary demand that we prove we are not bodiless brains connected to a supercomputer. The demand on its surface should be dismissed out of hand, as it is asking us to prove a negative. Why would anyone who is connected to reality ask us to prove such an absurdity? Answer: much of contemporary philosophy is not connected to reality, devoid of any understanding of Aristotle's primacy-of-existence premise.

[49] This excessive, rationalistic attention of traditional logic, and the absurd, brain-twisting p's and q's of what is called symbolic "logic"—scare quotes intended—is what has given rise to critical thinking courses in today's philosophy departments. Critical thinking as a concept is the attempt to bring reality and practicality back into logic, though the word "logic" is seldom used in those classes.

Hence, my often-made assertion that theoretical knowledge—science—consists mainly of concepts and principles, not concepts and propositions. Yes, all sciences use many concepts that are universal and, at the same time, make many statements that are not universal, but it is the universal principles that give science its power. The science of history is the only exception in that its goal is to discover and report concrete facts of a life or event.

The rules of formal logic, in effect, are the grammar and syntax of deductive reasoning, and non-contradictory identification applies, but inductive reasoning requires getting oneself out of ivory-towered discussions of language to look directly at reality.[50]

[50] On the argument that deductive reasoning adds no new knowledge because the conclusion of the syllogism is contained in its premises: this is only true of Scholastics and academics obsessing over whether Socrates is mortal, given that all men are mortal and Socrates is a man. Medical doctors certainly discover new knowledge about their patients when identifying the nature and causes of the patient's disease. Application of universals to identify a *this* as an instance of a *that* is new knowledge.

3

Related Issues

> To grasp a concept is to grasp and, in part, to retrace the process by which it was formed.
>
> —Ayn Rand[1]

> The same attribute cannot at the same time belong and not belong to the same subject and in the same respect.
>
> —Aristotle[2]

THE IMPORTANCE OF TRACING THE LINKS from broad abstractions to their roots in perceived concretes cannot be overemphasized.

This activity and, preferably, mental habit is not just a proper, updated Aristotelian empiricism that says, "Open your eyes and look at the world, then introspect to identify how you arrived at the concepts you have in your mind." It is a means of maintaining a non-contradictory mental organization that requires sharp focus on the facts of reality.

A significant side effect of this process is prevention of the previously mentioned mental habit called rationalism, whether of the Scholastic type or the more modern versions. Examples of the modern versions include the unreality of an economic theory that says capitalism must fit the mathematical formulas of a fantasy

---

[1] Ayn Rand, *Objectivist Epistemology*, 27.

[2] Aristotle, *Metaphysics*, 1005b19–20, in McKeon, *Basic Works*, 736.

called "pure and perfect competition," or the faux intellectual mistake that claims criminals by definition cannot be intelligent because they are irrational.[3]

Rationalism is a narrow focus on deductive reasoning at the expense of reality. It undermines the reality-focused deductive process of application that identifies "a this as an instance of a that."[4]

Thinking in well-constructed definitions that include a genus and differentia summarizing and condensing the content of the concept is how we keep our knowledge connected to reality and to the rest of the knowledge we already have. If, however, by "definition" we mean arbitrary and disconnected floating abstractions—the positivists' "arbitrary constructs"—knowledge can be anything we want it to be, which again is essentially where we are today as consequence of Immanuel Kant's philosophy and his positivist followers.

POSITIVISM

A significant point about Ayn Rand's theory of concepts is that it demonstrates how mistaken Kant and the positivists are. Abstractions from abstractions, such as space, time, and causality, and concepts of consciousness are not the innate, collectively subjective concepts of Immanuel Kant that are said to distort our awareness of true reality. As Kant concludes, we can only know the phenomenal world of appearances, never the noumena of "things in themselves." Nor are our broad abstractions "arbitrary constructs," as the positivists maintain—such as the "highly informative *guesses*" of Karl Popper and "wildly inaccurate" assumptions of economist Milton Friedman that we must test and attempt to "empirically verify," or "falsify," with an endless parade of statistical studies.[5] If any such abstract concepts

[3] Cf. Ayn Rand, "What Is Capitalism?" in *Capitalism: The Unknown Ideal* (New American Library, 1966), 7, for her assessment of the current state of economic theory.

[4] See my discussions of rationalism as a psycho-epistemology (or method of using one's mind) in *Montessori, Dewey, and Capitalism*, 74–76, 92–95, and *Independent Judgment*, 118n12. Ludwig von Mises uses the term "rationalism," but in the Enlightenment's sense of the supremacy of reason in science. On psycho-epistemology, see Nathaniel Branden, *The Psychology of Self-Esteem: A Revolutionary Approach to Self-Understanding that Launched a New Era in Modern Psychology* (1969; repr., Jossey-Bass, 2001), 96–98.

[5] Karl Popper, *Conjectures and Refutations: The Growth of Scientific Knowledge* (Routledge and Kegan Paul, 1963), 115. My italics. Milton Friedman, "The Methodology of Positive Economics," in *Essays in Positive Economics* (University of

cannot be so tested (or "falsified," according to Popper) they are declared meaningless and anyone who works with these abstractions is dealing in pseudoscience akin to astrology.[6]

The referents of "guess" and "hypothesis," however, which last Popper seems to assume is an informative guess, are not the same, which means they are not synonyms. Guesses have little or no evidence, while hypotheses have a significant amount, but not enough to draw a definite conclusion. Popper denied that he was a positivist, but his basic premises are positivist and falsification is just a variation on the positivists' verifiability principle.[7] The correct principle of truth, and of what is or is not valid science, is non-contradiction, a *mental* process of testing and trying ideas, requiring considerable cognitive effort, to demonstrate whether those ideas in our minds have correctly identified the facts.

The concept "positivism" that I am using is broader than the "logical positivism" of the Vienna Circle in the 1920s and '30s, which attempted, but failed, to find Kant's noumenal world by applying symbolic logic to sense data. Positivism, as it is still used today and derives from the work of August Comte, means the use of the so-called scientific methods of physics and chemistry to attempt to discover laws of physical, biological, and human events.[8] All science must be quantitative, and if it is not

---

Chicago Press, 1953), 14. Friedman, however, apparently repented writing this essay, Mani Basharzad, "Milton Friedman Regretted Writing 'The Methodology of Positive Economics,'" *Mises Institute*, December 27, 2024, mises.org; Deirdre Nansen McCloskey, *Beyond Positivism, Behaviorism, and Neoinstitutionalism in Economics* (University of Chicago Press, 2022), 60. On Friedman, see Kirkpatrick, *In Defense of Advertising*, 122–25.

[6] Walter Kaufmann mocked Immanuel Kant as "a virtuoso of rationalization" and quoted Popper critic, Adolf Grünbaum, saying Popper was "uninformed and unjust" when critiquing Sigmund Freud. Kaufmann, *Discovering the Mind: Goethe, Kant, and Hegel* (McGraw-Hill, 1980), 1:116. *Discovering the Mind: Freud versus Adler and Jung* (McGraw-Hill, 1980), 3:86. Kaufmann points out that German intellectual and statesman Johann von Goethe rejected the Newtonian notion that all science must be mathematical and "exact," advocating a more qualitative conception that included emotion, later influencing Freud. Kaufmann, *Goethe, Kant, and Hegel*, 35–42.

[7] Abraham Maslow, in *Toward a Psychology of Being* (1962; repr. Wilder Publications, 2011), 7–8, disapprovingly dubbed this approach to science as the experimental-positivistic-behavioristic methodology.

[8] "Positivism," *The Blackwell Dictionary of Twentieth-Century Social Thought* (Blackwell, 1993), 495–98. F. A. Hayek in his 1974 Nobel laureate address did not use the word "positivism," but did say that *scientism* and the *scientistic* attitude of insisting on using the methods of physical science in the social sciences "quite

quantitative, it is not science. The post-Kantian premises of positivists, in their attempt to defend science, are the following: metaphysics is meaningless—that is, the concept cannot be "empirically verified," and universals, facts, values, and, especially, truth are similarly unknowable. Awareness of reality is restricted to directly perceived concretes. At best, say the positivists, we can try to come up with those "successive approximations" of statistical probabilities, but we'll never reach "absolute" certainty or "eternal" truth, and, of course, all science must be "value-free."[9]

What this means is that positivism, which includes most of present-day science, does not only not have a theory of universals, it does not have a theory of abstraction from abstractions, let alone an explanation of how the process works. If one is a Kantian, which unfortunately economist Ludwig von Mises is to a minor extent, broad abstractions are said to be "a priori," not empirical, though they do apply (somehow) to reality. It is the positivists who dispense with Kant's innate categories and admit that their concepts are arbitrary "analytic truths," as they call them, unrelated to reality, with perceived concretes being the only things we can really know, called "synthetic truths."[10]

KANT'S INNATE CONCEPTS AND FREE WILL

Several additional issues require comment to distinguish Rand's views from that of contemporary philosophy. Three of Kant's allegedly innate broad concepts that impose order on our (phenomenal) world are space, time, and causality.

---

arbitrarily limits the facts which are to be admitted as possible causes of the events which occur in the real world." Hayek, "The Pretence of Knowledge," *Friedrich von Hayek Prize Lecture*, December 11, 1974, nobelprize.org.

[9] According to Rand, definitions, certainty, knowledge, and truth are not *changelessly* absolute, but in context they can be *contextually* absolute. Rand, *Objectivist Epistemology*, 47.

[10] See Leonard Peikoff, "The Analytic-Synthetic Dichotomy," in Rand, *Objectivist Epistemology*, 88–121. First published in *The Objectivist* 6, nos. 6–9 (1967). To emphasize, there is no "god's-eye," omniscient view of reality. Each one of us, like the blind men and the elephant, know only our portion, but this does not mean that extra effort exploring additional parts of reality will not generate valuable and true knowledge. Positivists, clinging to the mirror-of-nature theory of consciousness, seemingly want an effortless knowledge, ignoring and protesting the effort required to form and apply abstractions from abstractions and concepts of consciousness. (For stronger language, see Rand, *Objectivist Epistemology*, 67–68.) Aristotle talks about genus and species in his ladder of abstractions but does not have a theory per se of abstraction from abstractions. Aristotle, *Categories*, 1b10–25, in McKeon, *Basic Works*, 8.

Space and Time

Contrary to Kant, all these concepts have referents in perceptual reality. (Causality is addressed in the next section.) "Place" is a concept that is defined ostensively by pointing, so space is an empty place, such as a room that has no furniture (and is available potentially as an office or apartment), an empty or unoccupied parking location, or the spot on my desk where my water glass was. These examples are perceptual, metaphysical (out there) referents of the (in here) abstraction from abstractions of "space." Note that my examples can be described by further abstraction in geometrical terms, respectively, as three dimensional, two dimensional, or just a point, all perceptually grasped.[11]

Time, said Aristotle, is a "measure of motion" as the cubit is a measure of length.[12] Ayn Rand says that we form the concept of time from observing a change of relationships, for example, of an entity that moves from one place to another in relation to an entity that is stationary.[13] If I move my water glass from the right side of my computer to the left, I become aware not just of motion, but also of the passage of time. Aristotle defines time as "the number of movement in respect of the before and after."[14]

Historically, time was discovered as the changing phases of the moon in relation to the earth, then later as the movements of the sun across the earth's sky. It was measured initially by water clocks and sun dials. Today, we base our understanding of time on the revolutions of the earth around the sun and measure it with more and more precise

[11] Cf. Aristotle, *Physics*, 212a5–31, in McKeon, *Basic Works*, 277–78, who preferred to use "place" instead of "space." Aristotle regarded mathematical concepts as abstractions from abstractions, stating that the mathematician "strips off all the sensible qualities [of perceptual concretes] . . . and leaves only the quantitative and continuous, sometimes in one, sometimes in two, sometimes in three dimensions." *Metaphysics*, 1061a30–35, in McKeon, 855. By implication, using Rand's theory, we first perceive the solid object, then abstract from it the concept of plane, then from plane, the concept of line, and finally from line, the concept of point, which is the highest-level abstraction. Cf. *Topics*, 141b10–11: "A solid falls under perception most of all, and a plane more than a line, and a line more than a point." Barnes, *Complete Works*, 1:239.

[12] Aristotle, *Physics*, 221a1–3, in McKeon, 294. A little later, 221b10, Aristotle says time is "not motion, but number of motion," 295. Cubit is an ancient measure of length that extends from one's bent elbow to the tip of the middle finger.

[13] Rand, *Objectivist Epistemology*, 258–60.

[14] Aristotle, *Physics*, 220a25, in McKeon, 293.

time pieces. Because we must know several prior concepts—entity, change, motion—time is a broad abstraction whose referents can be traced back to perceptual concretes. It is not innate.

Causality

Causality, according to Rand, is decidedly an abstraction from abstractions formed from and traceable back to perceptual reality. It is not an innate Kantian category.

For Rand, causality is "the law of identity applied to action."[15] It is the actions of one or more entities with their attributes in relation to the actions of one or more other entities with their attributes. Essentially, this is Aristotle's formal cause, including his distinction between potentiality and actuality, and a rejection of the view that has dominated philosophy of science since at least the time of Galileo and Newton. That view looks only at efficient causation, the so-called billiard-ball causality. One of Rand's fundamental propositions is that an entity is *all its attributes* and that is its identity. There is no metaphysical substratum or glue holding the attributes together, which would take us back to the intrinsic theory of essences. Thus, to arrive at a causal explanation of an event we have to recognize the nature of the entities involved.

A billiard ball, for example, going in the pocket of a pool table is not fully explained by saying the cue stick moved in a certain way to knock it in (efficient causation). We have to know that the balls are hard and roll easily on the fabric of the table and a well-chalked cue stick in the hands of a skilled player with good vision hits the ball at the right angle and speed. The concepts of a round and hard ball, smooth and flat table, chalk and cue stick, and skilled player with accurate eyesight all identify attributes of the respective entities and their interactions to cause this event. Simple billiard-ball causation is not so simple. As can be seen, Aristotle's other three causes—material, final (when talking about human and other living action), and efficient are relevant in a full explanation of a cause. But formal cause, the nature or identity of the entities involved, is central, and it subsumes the material, final, and efficient causes. They are all aspects of identity.

The implication of Rand's view that causality is identity in action is that essentialization is a grasp of causality. Conceptualizing through essentialization identifies causes and effects of the existents that the

---

[15] Rand, *Atlas Shrugged*, 1037.

concept identifies. The essential distinguishing characteristic of an entity and Rand's rule of fundamentality say that to identify the essential characteristic of a concept, we must identify the one or more characteristic(s) that explains and causes all or most of the others. Again, this means that the explanation is "in here," in our internal mental process of consciousness and is epistemological, and the causal relationship is "out there," in reality apart from that mental process and is metaphysical.

Thus, cholera is explained and caused by the essential distinguishing characteristics of the comma bacillus (today called *Vibrio cholerae*) interacting with the digestive system of the human body. Dew is explained and caused by the characteristics of water condensation, that is, of water vapor in the air interacting with air temperature such that liquid forms on our cars, windows, and leaves of grass. And tides are explained and caused by the attractions between water on earth to the sun and moon as they move, especially the gravitational pull of the moon on the oceans of earth; those oceans that are on the side of earth closest to the moon, and furthest away, "bulge out" and create high tides. The in-between oceans, depending on rotation of the earth, experience low tides.

Which is not to say that the above explanations are exhaustive of the respective causes. Qualifications are often required, as the "simple" explanation that water boils at 212° Fahrenheit requires the qualifications "varying by air pressure and purity of water." Nevertheless, the many concepts involved, which means the many entities with their specific attributes involved, had to be examined in detail through testing and trying in various experiments to arrive at the final essential distinguishing characteristics. And the word "final" must be taken advisedly as this does not mean these findings are true "eternally." Knowledge grows and discoveries increase, meaning that our causal definitions are contextual and may need to be adjusted. Einstein's theory of gravity in relation to Newton's is just one example of this "editing" of a previous theory, in the light of greater knowledge.

Conceptualization by measurement omission, which identifies the cause of an event by singling out the essential distinguishing characteristic or characteristics of the existents involved, is the essence of theory in both basic and applied sciences. Measurement is not the essence; it is an *aid*, which may be extensive in some sciences, to the discovery and application of theory.

An appropriate note here is to point out that many controlled experiments performed today aimed at determining cause and effect relationships are superfluous, correlational, or performed with less than sound methodology, such as inadequate assumptions or insufficient study time to identify accurate effects. Many such studies only generate historical data, not theory.[16] Contemporary psychoanalyst Jonathan Shedler states that we do not need to conduct RCTs—randomly controlled trials, as they are called in some sciences, such as medicine—to know that "the sun causes sunburn, sex causes pregnancy, or food deprivation leads to starvation."[17]

These examples, Shedler continues, are known by observation because we know their mechanisms (or methods) of action. And these examples, I would add, are applications of well-known concepts that illustrate Aristotle's formal cause and Rand's theory of causality. "Mechanism of action" means that by identifying the entities and their attributes in a causal situation—sun, skin, and sunburn; sex organs, sex, and pregnancy; nutritional organs, food, and starvation—we can know their actions and effects on each other. Conceptualization, grasping the essential distinguishing characteristics of the entities and attributes involved, which includes retention of all the knowledge we have learned to date about the entities, that is, the information in our respective file folders, are key to knowing the operation of any action.[18]

Free Will

A key application of causality as identity in action is that the essential distinguishing characteristic of homo sapiens is the animal that possesses the capacity to reason. The word "capacity" means we do not

---

[16] Or, as I wrote in *In Defense of Advertising*, 163n50, much research today in the human sciences "merely verifies the obvious and belabors the trivial."

[17] Jonathan Shedler, "Selling Bad Therapy to Trauma Victims," *Psychology Today*, November 19, 2017, psychologytoday.com. Cf. John P. A. Ioannidis, "Why Most Published Research Findings Are False," *PLoS Medicine* 2, no. 8 (2005): 696–701, https://doi.org/10.1371/journal.pmed.0020124, and Ioannidis, "Why Most Clinical Research Is Not Useful," *PloS Medicine* 13, no. 6 (2016): 1–10, e1002049. https://doi.org/10.1371/journal.pmed.1002049.

[18] "Mechanism of action" is a technical phrase used in pharmacology at the molecular level to explain the effects of a drug on the human body. I have generalized it to the epistemological level. I prefer the words "method" or "process" of action to avoid concessions to materialism. Jonathan Shedler, in "It's Time for Psychology to Lead, Not Follow," *Psychology Today*, February 25, 2025, psychologytoday.com, rejects materialism using an interesting analogy: the movie we are watching on television supervenes, that is, is more than, not the same as, the pixels generating the movie. Similarly, he adds, "Neurobiology is one level of analysis, mental life is another," which leads him to emphasize in italics, "Psychology should be like psychology"—not medicine or any other physiological discipline.

reason automatically. We must choose to do so. This is the meaning of free will or volition, another notion that Kant cannot find or explain empirically. Kant's phenomenal world is the world of science, which is the world of determinism. Thus, free will for Kant must reside in that (unknowable) noumenal reality. How does he know? He says free will (and God and immortality) are necessary for moral responsibility. This is wishing, not rational argumentation.[19]

Human free will is a volitional consciousness that requires cognitive self-regulation to avoid errors.[20] It is a choice to think or not to think to perceive reality correctly by formulating accurate concepts that will guide our choices and actions to achieve a happy, healthy life. Failing to think, or to think accurately, leads to failure and unhappiness.

How do we know we have free will? Leonard Peikoff states that "it is a self-evident fact, available to any act of introspection."[21] This, again, is the Aristotelian notion of self-evidence as directly perceived. Further, any attempt to deny free will as self-evident requires the acceptance of it. Determinism, the denial of free will, says we have no real alternatives in our lives, because choice is an illusion caused by genes, environment, or other unseen forces. This is a self-contradiction. Determinists are presumably determined to believe in determinism, yet they assume they are logically controlling their own minds. The many dodges and equivocations they use to get around this contradiction do not justify their doctrine.

Free will, as Peikoff says, is a corollary of the axiomatic concept of consciousness.[22] Since consciousness is a fallible natural attribute of human beings—not a supernatural gift from God—causality for us means we have the *potential* to choose to exert effort to acquire those concepts that will guide us, including those abstractions from abstractions, by following the laws of logic to create the guiding knowledge. This means we must *actualize* that potential by exercising due diligence when using our minds. We are the cause of our thoughts and actions.

---

[19] Hegel looked at Kant's case for an unknowable noumenal reality and concluded no such thing exists, only the phenomenal.

[20] See Harry Binswanger, "Volition as Cognitive Self-Regulation," *Organizational Behavior and Human Decision Processes* 50, no. 2 (1991): 154–78, https://doi.org/10.1016/0749-5978(91)90019-P.

[21] Peikoff, *Objectivism*, 70.

[22] Peikoff, 71.

The nature of homo sapiens and its consciousness is the formal cause of human action. It is us—each one of us—who must decide what to do with our lives and how to live it.

I mentioned in chapter 2 that the higher animals may have a modicum of choice.[23] The family dog, for example, has a sophisticated perceptual system that enables him to sniff and taste to know what would be good for him. Animals, though, like humans do make mistakes. He just does not have what we call a *free* will. The dog's perceptual system enables him to eat the food set before him, or not, even though it may be his favorite from days before. He may choose to go into the other room, to take a nap, and so on. Clearly, the higher animals cannot view existents as units to form concepts, because they are perception-bound, but they do have some amount of control over their lives. Free will is not an exception to causality. It is a type of causation.[24] The nature of the entity—dog or human—is what determines its actions, but those actions are determined by the attributes of the entity: limited choice or human free will.[25]

The will of humans refers to the mental process of selecting and directing our thoughts, choices, and actions. The will of the family dog is its power or ability to select an action within its range of perceptual possibilities. In the context of evolution, I would hypothesize that

---

[23] See above, p. 43. See also below, p. 97.

[24] Peikoff, 55–72. Rand says that "volition begins with the first syllogism," conceptualization, but this ignores the adventurous curiosity of pre-verbal (pre-conceptual) toddlers who *choose* to go into this room, then into that one, then to pick up a handful of sand to see what it tastes like. Such behavior is similar to the higher animals, such as the family dog. Ayn Rand, "For the New Intellectual," in *For the New Intellectual: The Philosophy of Ayn Rand* (Signet Books, 1961), 15.

[25] Mortimer J. Adler, in *The Difference of Man and the Difference It Makes* (World Publishing, 1968) 153–54, suggests that the higher animals possess the ability to abstract perceptually—perceptual abstraction, as he calls it—by recognizing, and to some extent, remembering their awareness of sameness and similarity in perceived concretes. If explored further, this notion would be a step forward in acknowledging the scientific, biological evolution of free will as something that did not come from God. Concept formation animals cannot do, but emotional generalization that we observe in young children can also be observed in the higher animals. Cf. Branden, *Psychology of Self-Esteem*, 30–31, and Kirkpatrick, *Independent Judgment*, 67–68. In the latter work, I elaborate Edith Packer's concept of emotional generalization as beginning when we are quite young, rooted in our experiences of pleasure and pain. Higher animals react similarly, say, for example, when the family dog experienced mean treatment from a previous owner who was a big man. In the present he may express fear of not just the present owner but of big men in general.

the essential meaning of consciousness, as possessed by humans and some lower species, is, in addition to being an awareness of the species' surroundings, the ability to select from alternatives to guide its actions, to sustain its life. The genus of this biological concept of consciousness would be the means or method of responding to the environment and the differentia would be the various organisms' methods, such as sensation only, sensation and perception only, or the humans' experience of sensation, perception, and conceptualization via the exercise of reason. Rand does not discuss choice in the lower species.[26]

Defensive Habits and Free Will

Ayn Rand describes human volition as the choice to think or not to think, that is, to focus our minds to direct and control attention to something in particular, such as a problem to be solved in our work, or to let our minds wander away in daydreams or a daze. To focus means to maintain "a full, active, purposefully directed awareness of reality,"[27] to exercise our reasoning capacity in the process of acquiring knowledge without contradiction and in keeping our subconscious minds well-ordered to guide our lives in moral and successful ways. Focus means maintaining an active mind even in leisure activities. The decision to think or not, however, especially the quality of thinking we generate, can face obstacles and is not always an issue of volitional control.

Defensive habits are significant obstacles to clear thinking. Defensive symptoms of the neurotic type are a form of delusion—though not nearly as serious as the delusions of a person experiencing a psychotic episode. A young man, for example, who is fired from his job and jilted by his lover on the same day may become seriously depressed and conclude, "I'll never find another job or lover."[28] This is not true and can be called a delusion about reality. This person,

[26] See neuroscientist Antonio Damasio, *Feeling and Knowing: Making Minds Conscious* (Vintage Books, 2021), 17, 109–110, and throughout, for his discussion of the four means by which a species responds to its environment, consciousness being the highest. Sensing or detecting (the most basic form in any organism), minding (beginnings of a nervous system in animals), and feelings ("hybrids of mind and body") are the other three. Selection from among alternatives, I submit, which must have arisen at some point in the evolutionary process, is the root of human free will. In other words, our free will did not spring sui generis from the brain of homo sapiens.

[27] Rand, *"Objectivist Ethics,"* 24.

[28] I used this example in *Independent Judgment*, 66–67.

in addition, may, without help, have considerable difficulty doing much of anything for several days or weeks. His thought processes are turned off, partially or completely.

Much of what we think, feel, and do as adults has been influenced and shaped by the conclusions we make as children and teenagers. An influential part of our psychologies is what Edith Packer calls core evaluations about ourselves, other people, and the world in general.[29] How well these premises have been formed, meaning how correct and healthy they are, strongly influences how we will act later. This gives rise to a question: how well can we focus our minds to develop a well-ordered subconscious when we experience a host of psychological problems? Not easily is the answer. The formation of these premises depends greatly on our parental upbringing and schooling. What we believe and feel as adults is not as simple as flipping a switch, though generally, absent drug influence or physiological damage to our brains, behavior is controllable.[30] Which means we can refrain from pulling out a gun and shooting someone, or cheating another person through a dishonest act, but certain areas of our lives may still not exhibit what an outside observer would call clear thinking.

The influence of defensive habits, I believe, is underemphasized in Ayn Rand's writings, though she does call these types of failures to focus and think errors in knowledge, as opposed to willful evasions.[31] Technically, this is true. The person with psychological problems does have free will and did in the younger years when initially creating the false premises, though the formation of many of these premises occurs by emotional generalization, even chance.[32] Many of the false premises have been repressed and shaped into habits manifested as symptoms that cloud our perception of reality. Such a person often does

---

[29] Packer, *Lectures on Psychology*, chap. 1.

[30] Peikoff, *Objectivism*, 58–59, says Rand considered the act of focusing to be like throwing a switch. I do not think the analogy is accurate about our mental processing, especially when we consider psychological problems. To view focus and thought as throwing a switch is to ignore psychological issues.

[31] Shooting or cheating someone is clear evasion of what is right or moral. The criminal personality is one who lies, cheats, and steals as a way of life and enjoys getting away with the forbidden, primarily "justified" by the defensive habit of rationalization. See Stanton Samenow, *Inside the Criminal Mind*, rev. ed. (1984; repr., Broadway Books, 2014), esp. chap. 12.

[32] On chance emotional generalizations, see Kirkpatrick, *Independent Judgment*, 68n29, and "Chance Emotional Generalizations and Introspection," *Jerry Kirkpatrick's Blog*, June 7, 2019, jerrykirkpatrick.blogspot.com.

not know how to correct the errors. And with the present influential view of determinism by genes and environment, many unfortunately conclude, "That's me and I can't do anything about it." However, free will as controlling behavior means we can seek professional help or continue to live our lives despite the obstacles.[33]

Dealing with our psychological problems is more difficult, especially in today's culture, than many realize.[34] Failing to understand the role of psychological problems in personality development can lead to what I call moralism or a moralistic attitude, a tendency to attribute volitional control to nearly every thought we have and act we take, with consequent tendency in the extreme case to condemn as immoral nearly every thought and behavior different from our own. Moralism results from a failure to understand how powerfully subconscious premises, dating from early childhood, can influence present thought and behavior, in some cases, preventing us from knowing why we think what we think and do what we do. Psychology and philosophy, especially ethics, are distinct sciences; they each have appropriate contexts in which they apply.

KNOWLEDGE AND EMOTION

Two issues remain: that knowledge is not the monopoly of the highly educated, deposited primarily in publications, and that emotions are present in every act of knowing and doing, including in science and scientists.

General versus Personal Knowledge

Several times in previous pages I have made reference to our personal, everyday knowledge and how the layperson uses concepts and universals. Personal knowledge should be distinguished from general knowledge and I first made the distinction in *Independent Judgment and Introspection*.[35] Here are two quotations from that work:

---

[33] Talking to a good friend may help, though such help usually is empathy rather than sound therapy.

[34] Alcoholics who want to stay sober must every day confront strong urges for a drink. Clouded thinking is often the result. The same is true of neurotic symptoms. Jonathan Shedler, in "That Was Then, This Is Now: Psychoanalytic Psychotherapy for the Rest of Us," *Contemporary Psychoanalysis* 58, nos. 2–3 (2022), 432, https://doi.org/10.1080/00107530.2022.2149038, points out that sound, effective therapy "*creates* freedom, choice, and a freer will," 432. Shedler's italics.

[35] Rand did not deal with this distinction, at least not in any length.

> To survive and flourish, humans require general knowledge, or education, in the form of concepts and principles to guide their choices and actions. Each individual, in addition, requires specific knowledge, or a set of concepts and principles unique to his or her experience, to direct action to the achievement of health and happiness.[36]

> Systematic bodies of concepts and principles constitute our sciences; specific bodies of concepts and principles, unique to our own experiences, constitute our personal knowledge.[37]

General knowledge derives largely from others, essentially our education broadly construed, not just what we learn in school. For many, unfortunately, most of that general learning ends in the last year of schooling. Personal knowledge, which includes the general, is what we have experienced over the years and makes us who we are as individuals.

Personal knowledge comes before the general, as we saw earlier with the infant's epistemology of experiencing sensations and perceptions, before the toddler's and young child's formation of first concepts. Personal knowledge starts to develop in reaction to pleasureful and painful experiences. The pleasure-pain aspects of these experiences provide the foundation of our emotions, and the emotions we associate with concrete experiences in childhood and youth influence how healthy and happy we will be as adults.

Personal knowledge, as a result, might also be called experiential knowledge, the acquisition of knowledge based on the objects, persons, events, and even ideas we come in contact with or, rather, experience firsthand. Personal knowledge is a collection of concrete experiences—with parents and other relatives; with school subjects, teachers, and classmates; of the time we learned how to ride a bicycle; of happy and sad times at summer camp; and, as an adult, of work, family, and leisure. Throughout life. As concrete experiences, personal knowledge, to bring back the theory-history distinction, is our (personal) history. The general knowledge we have acquired is theory.

General knowledge becomes part of our personal knowledge because it is what we have learned and is a part of us, and it may or may not be individualized to our interests and goals. Personal knowledge forms the basis of our personalities, our distinctive ways

[36] Kirkpatrick, *Independent Judgment*, 54.

[37] Kirkpatrick, 73.

of thinking and acting. Much of our personal knowledge is specific to concrete objects, persons, and events, such as a ball bouncing and rolling seen as a young child or, as an adult, a recent performance of a Brahms symphony and maybe the pleasant (or unpleasant) experience of changing a tire on the family car.

Personal knowledge is not universal in the way general knowledge is, though it can become general if, say, we continue to increase our general knowledge, thus enabling us to write an article or book about Brahms symphonies or the repair of automobiles. Most of what we do with our personal knowledge is apply the general we have learned previously to individual experiences.[38]

The application of knowledge, whether general or personal, we must now emphasize, is not an emotion-free process. This takes us to the place of emotions in the formation of universal concepts in both everyday life and in science.

The Place of Emotions in Science

Thoughts underly our emotions. Psychologist Edith Packer points out that when we are feeling something, we also are thinking something.[39] The opposite also is true—when we are thinking something, we are feeling something, which means we have emotions running through our minds whether we are aware of them or not. Consciousness, to borrow a word from William James, is a stream, not separate unrelated compartments that can be shut off or turned on at will, though the evaluations that stand behind emotions can be repressed. Scientists clearly have emotions, not just in their eureka moments of a major discovery, but in their typical workdays. Emotions, after all, are what motivate us.

---

[38] The notion of personal knowledge is what F. A. Hayek refers to when he denies that central economic planning can ever succeed, because such planning essentially assumes omniscience on the part of whoever is making the planning decisions. Because none of us is omniscient—we each only know our small part of reality, as do the blind men about the elephant—our personal knowledge is what drives the marketplace of capitalism, with the price system coordinating all our individual, *personal plans*, not what we, or rather the central planners, learned in graduate school. Friedrich A. Hayek, "The Use of Knowledge in Society," in *Individualism and Economic Order* (Gateway Editions, 1948), chap. 4. See also George Reisman, *Capitalism: A Treatise on Economics*, (Jameson Books, 1996), 137–39.

[39] Packer, *Lectures on Psychology*, 278.

When we acknowledge the presence of emotions in science, and scientists, we acknowledge that evaluations are present. Behind every emotion, as Packer also points out, are two evaluations, one universal that applies to all experiences of a particular emotion, such as joy or anger, and a personal one that includes all the concrete details of the moment when we feel it. Joy, for example, at the universal level means "I have achieved an important value." Thus, "eureka"—or "I have found it," the translation of the Greek—is a form of joy. The scientist's personal evaluation—"all of my years of research have been worth it," for example—in the eureka moment might be expressed in his behavior as energetic shouting and jumping up and down such that he almost knocks a beaker off the table. This would be a personal evaluation and experience that likely would last a long time in the scientist's subconscious memories.[40]

It is these inner conversations or voice, as Packer describes personal evaluations, that constitute a key part of the content of our personal knowledge. Which means that knowledge, general or personal, is not "value-free" as the positivists insist it must be.[41] Every emotion exhibits not just an evaluation, but also an action tendency, or urge to act. The scientist who, when young, is given a chemistry set for a birthday may later get excited in the high school chemistry class and begin to think about a career in the field. Meeting and talking to professional chemists—a chemical engineer and a research chemist, for example—may help the young person solidify his or her career goals. Going to work every day in an office or lab is motivated by our emotions of pleasure or pain we associate with the work.

Emotion interacts with thought to drive everything we do. Writers of both fact and fiction say they follow their emotions to come up with subjects, themes, and even the phrasing of sentences. Many write "by ear," more so than by the current rules of grammar and syntax.[42]

---

[40] Another example: the universal evaluation of anger says, "An injustice has been done to me," and a personal evaluation might be: "a colleague whom I had respected stole my idea and published it without permission or acknowledgment." See Packer, *Lectures on Psychology*, 111–12, 226–35.

[41] See chap. 5, pp. 133–140, for more on values and "value-free."

[42] I have told students who could not come up with a term paper topic to "go with what grabs you," meaning what your emotions are telling you. See davidkubicekwriter.com on writing by ear. Cf. Ayn Rand, *The Art of Non-Fiction: A Guide for Writers and Readers*, ed. Robert Mayhew (Penguin, 2001).

Whatever we have liked or disliked in the past, influences our present. Emotions contribute knowledge to our thought processes.[43] Those past emotions often are the sources of connections we make in the present, sometimes called creative insight, as well as guidance to follow particular lines of thought and experimentation.[44]

Repressed persons are also motivated by past emotions, though usually negative ones that they seek to avoid. Some who are repressed may appear in the present to experience no emotions. A severely repressed scientist, for example, may exhibit a highly muted reaction to the eureka moment, such as a sober, unexpressive face and words that say, "This is good." But the emotions, which means evaluations and values, are there at some level. Such a person is often desperately trying to avoid the error and appearance of the fallacy of the appeal to emotion.

Where does this fallacy come into play? It is actually a simple notion that says it makes no sense to say or write, "It is true because I feel it." This does not mean that we should not be motivated to become a scientist because of our emotions from the past or that we should not get excited over a eureka moment. It is unfortunate that the repressed scientist does not jump up and down.

The answer to the question about the appeal to emotion is that whatever emotions we have in the process of identifying a fact of reality, we must follow Aristotle's principle of non-contradiction. Thus, in our work, we bring up our emotions to see if any of them are undermining our perceptions of reality. If I am feeling that a leprechaun is on my desk instead of a glass of water, I assure you I am committing the fallacy. Or, to bring back the young man fired and jilted on the same day, he *feels* that he will never find another job or lover. This is all the fallacy refers to, a feeling (in here) that contradicts the facts (out there).

---

[43] See a more in-depth treatment of this issue in Linda Reardan, *Emotions: A Modern Aristotelian View* (Kirkpatrick Books, 2026), chap. 5, such as: "Once the errors that are contained in many emotions are factored out, what is left is action-tendencies that are anything but arbitrary: They push us to pursue unimpeded life-activities and to avoid impeded ones, in the same way that physical pleasure and pain do, except that emotions give us information not about merely bodily needs, but about the needs of consciousness," Reardan, 125.

[44] For example, when researching *Montessori, Dewey, and Capitalism*, I was surprised to find connections and insight, after reading them, between Maria Montessori, John Dewey, and Ayn Rand. They obviously disagree about many things but nevertheless exhibit some key similarities.

Thus, scientists, I am quite certain, when coming up with a solution to a problem can, without guilt, go wild and crazy to celebrate![45]

UNIVERSALS AND CAPITALISM

The significance of a correct theory of universals is that our knowledge consists of universal concepts. An invalid theory of concepts undermines everything else we can know. A correct theory allows our knowledge to grow and thrive, thereby guiding us to make wise choices and take rational actions.

To develop and defend a philosophy of capitalism, we must be able in particular to validate the concept of value to establish a theory of ethics, and we must then be able to apply the concepts of ethics to political philosophy to establish valid concepts of government, rights, and freedom, and to further apply all of these previously formed concepts to societal institutions.

To develop and defend a psychology of individualism and an economics of capitalism, we must be able to validate all the relevant concepts in these two fields. We turn now to the two best representatives of, and originators in, psychology and economics, Sigmund Freud and Ludwig von Mises.

[45] "The self-denial of those scholars," Walter Kaufmann says in *Discovering the Mind: Nietzsche, Heidegger, and Buber* (McGraw-Hill, 1980), 2:10, "who do not permit themselves any strong emotions, such as powerful admiration and detestation," or use of the word "I" in their writing, "deprives them of an important aid to self-knowledge." Such "lack of self-awareness," Kaufmann continues, "is frequently mistaken for objectivity." This is another example of the fear of emotion in science.

II. Freud and Mises

4

Sigmund Freud as Scientist

> This is the way of science: slow, groping, laborious. . . . Through observation one learns something new, now here, now there, and at first the pieces do not fit together. One formulates surmises and makes auxiliary constructions that one takes back when they are not confirmed . . . in the end all this exertion proves worthwhile, the scattered finds do fit together, one gains an insight.
>
> —Sigmund Freud[1]

> I remain a liberal of the old school.
>
> —Sigmund Freud[2]

ANY THEORY OF SOCIETY—free or otherwise—rests on a theory of human nature.

Psychology as a basic science is a former branch of philosophy that defines such a theory. Philosophical psychology identifies the nature and role of consciousness in human life. Based on this theory, psychology as we know it today identifies concepts and principles that explain human motivation and behavior. In its applied version,

[1] Sigmund Freud, *New Introductory Lectures on Psycho-Analysis*, trans. and ed. James Strachey (1962; repr., Norton, 1989), 215–16. First published in German in 1933. Epigraph translation by Walter Kaufmann in *Freud versus Adler and Jung*, 79–80. Subsequent "first published in" references to Freud's works should be assumed to be the original works in his native German.

[2] Freud to Arnold Zweig, November 26, 1930, in *The Letters of Sigmund Freud and Arnold Zweig*, ed. Ernst L. Freud, trans. Elaine and William Robson-Scott (Harcourt Brace, 1970), 21–22.

psychotherapy, it provides concepts and principles to guide professionals to help patients live mentally healthier and happier lives—and to help laypersons introspect to correct unhappy mistakes. Based on the conclusions it draws about human nature, psychology lays the foundation for recommending how individuals can and should interact in a social setting—free or otherwise.

Sigmund Freud is the father of modern psychology.[3] Almost single handedly, he challenged the prevalent notion of his time that psychological symptoms were "nervous diseases" caused by physiology. Though he was trained in neurophysiology, his experience with patients in practice led him to identify the basic concepts of psychology as a *mental* science and to identify the methods of treating psychological problems and their symptoms by working out the *mental* causes with his patients.

Let me now review Freud's accomplishments and contributions to psychology as understood in the context of Ayn Rand's theory of concepts. In chapter 5, I will do the same for Austrian economist Ludwig von Mises. The aim of this chapter and the next is to demonstrate how science works in the arena of defining concepts and principles about human beings who possess a volitional consciousness. In the process, I will be demonstrating how both Freud and Mises are true scientists.

The issue of volitional consciousness, or free will, however, becomes problematic for both Freud and Mises and will be addressed as it comes up. The main question is, can humans change themselves or is their behavior determined by genes and environment? Freud and Mises can be described as "compatibilists," a recent and bizarre contradiction used today that says free will and determinism are compatible, meaning we can act *as if* we have choice and control, including moral control, over our lives, even though determinism is said to be true. Compatibilism goes hand in hand with complacent skepticism, which today means it goes hand in hand with "good-enough" science.[4] A final assessment in this chapter will address Freud's determinism, which I think is his biggest failing.

---

[3] Aristotle is the father of psychology.

[4] Which is to say "good-enough" science today is not good enough, for example, to send human beings safely in rockets to the moon or Mars—and back. Nor is it "good enough" to enable people to overcome mental problems or to establish and keep a free society. Certainty within a context, not glossed-over contradictions, nor omniscience or eternal knowledge, is prerequisite for sound science.

FREUD'S EPISTEMOLOGY

Sigmund Freud is a peerless example of both a basic and applied scientist of human nature, of the human science of psychology. Psychoanalysis is his basic theory, while psychoanalytic therapy is the applied science or practice of the basic theory. Freud called psychoanalysis the "science of the unconscious," or depth psychology, while psychology in general, according to Freud, is the science of the soul, which consists of three parts: our conscious reasoning mind and seat of emotions (*the I*), our conscience (*the upper I*), and our unconscious (*the it*). Freud was an atheist his entire life, as well as a determinist, so the soul for him is not immortal nor does it exhibit a free will.[5]

Freud as scientist, we might say, is a post-Kantian, modern Aristotelian. "Post-Kantian" means that consciousness is recognized as active and creative, not passive or mirror-like, while "modern Aristotelian" means recognizing that an active, creative consciousness can, guided by Aristotle's logic, accurately perceive reality. A modern Aristotelian in this sense rejects Aristotle's moderate realism that the essence of a thing is "out there" in the thing but retains his premise of the primacy of existence. Consciousness does not create or distort reality, as Kant and other philosophers say, or reflect or mirror it, as the moderate realists think. The modern Aristotelian holds that consciousness *creates*—using the word "create" in the epistemological sense—concepts and principles based on what it has correctly recognized or identified of reality.

As a student at the University of Vienna, Freud attended the lectures of Aristotelian scholar Franz Brentano and was influenced by him.[6] In an 1875 letter to Eduard Silberstein, Freud writes about several visits, along

[5] See the next section of this chapter for my discussion of the challenge of reading Freud in general and in English translation in particular. When quoting Freud or referring directly to his work, I will continue to use his concept of the unconscious, though sometimes I will write the more modern term "subconscious." In *Independent Judgment*, 177–81, I explain why I prefer the concept "subconscious"—mainly because we are talking about a lower level of awareness, which occurs even in sleep. We are unconscious of many things, the circulation of our own blood, for example, and most of the rest of the universe, but what we repress is not truly unknown or forgotten.

[6] Franz Brentano, *Psychology from an Empirical Standpoint*, trans. Antos C. Rancurello, D. B. Terrell, and Linda L. McAlister (Routledge, 1973). First published in German in 1874. In this work, Brentano frequently uses the word "phenomena" (*Phänomene*) when talking about the objects of awareness, putting him in the neo-Kantian tradition. In a later essay, included as an appendix to *Psychology from an Empirical Standpoint*, 321–30, "On Objects of Thought," 1915, he asserts that we can directly know *Reales*, the real thing. Cf. Hamid Taieb, "Intentionality

with classmate Joseph Paneth, to Brentano's home. The purpose of the visits was primarily for the professor to convince the two materialistic doubters of the existence of God.[7] The Aristotelian influence on Freud can in particular be seen in his frequent use of the words "nature and essence," often followed by "origin" (or cause) of a mental process that he was trying to identify. He never went looking for Kantian noumena or intrinsic essences or Newtonian algebraic equations, though he does recognize that both normal and abnormal psychologies exhibit quantities of emotional energy measured approximately on an ordinal or qualitative scale of less and more.[8]

Freud rarely mentions Kant in his writings, but does, despite his early interest in becoming a philosopher, make uncomplimentary remarks about the philosophers, most of whom in his time were neo-Kantians. In the same letter to Silberstein, Freud agreeingly relates Brentano's opinions of several German philosophers. For example, Kant—Brentano's words as reported by Freud—is "an intolerable pedant" and his successors, Schelling, Fichte, and Hegel, are to be dismissed "as swindlers." Also, Herbart's "a priori constructions" were condemned, because Brentano, wrote Freud, was "unreservedly a follower of the empiricist school."[9]

---

and Reference: A Brentanian Distinction," *The Monist* 100, no. 1 (2017), 120–132, https://doi.org/10.1093/monist/onw022. Brentano and other philosophers thought an "unconscious" consciousness was a contradiction in terms. Brentano, 102–105. See one of Freud's responses to his critics of the unconscious in *The Ego and the Id*, trans. Joan Riviere, rev. and ed. James Strachey (1960; repr. Norton, 1989), chap. 1. First published in 1923.

[7] Freud to Eduard Silberstein, March 15, 1875, in *The Letters of Sigmund Freud to Eduard Silberstein 1871–1881*, ed. Walter Boehlich, trans. Arnold J. Pomerans (Belknap Press, 1990), 102–104. See also "A Diary or Chronology of Important Events in Freud's Life," Richard G. Klein, *Freud 2 Lacan*, freud2lacan.com. The timeline covers Freud's early years to 1900. Freud briefly considered Brentano's arguments about God but was not convinced. He and Paneth at the time referred to themselves as materialists.

[8] This is Freud's *economic* perspective on consciousness, "quantities of excitation," as he refers to them. The *topological* perspective is his three components of the soul, called the conscious, the preconscious, and the unconscious, which later became his "structural" conception: *the I*, *the it*, and the *upper-I*. The *dynamic* perspective refers to the nature of the unconscious, gradually formed from birth, as having the power to influence consciousness, especially when we are not aware of such activity.

[9] Freud to Silberstein, March 15, 1875, 102, 104. Kemp Smith, in his *Commentary*, xix, xx, writes that Kant's *Critique of Pure Reason* was "hastily thrown together" and that "Kant flatly contradicts himself in almost every chapter." As a writer, Kemp Smith adds, Kant "is the least exact of all the great thinkers."

In his own writings Freud several times quotes German poet Heinrich Heine's comment on the typical German philosopher:

> *With his bathrobe rags and long nightcaps*
> *He mends the world and plugs all the gaps.*

According to Walter Kaufman, Kant and Hegel wrote their masterpieces in bathrobes and nightcaps, while plugging the gaps to complete their rationalistic systems.[10] Elsewhere, Freud expresses disapproval of the "forced and tortuous expressions" of German and Austrian intellectuals.[11]

As I pointed out in chapter 1, Freud spent many years working out the correct meaning of repression as distinguished from the more general concept of defense, clarifying the distinction in 1926.[12] In those earlier years he had considered the two terms synonymous. He recognized finally that repression had to be narrowed in its definition as it is a defense that responds to anxiety by attempting to allay or quash it; it is not a cause of anxiety, as he had earlier thought. Neurotic anxiety is what gives rise to the need for repression.[13]

This is conceptualization, and how science works.

As Freud wrote in *An Outline of Psychoanalysis,* "All sciences are based on observation and experience that are mediated by our psychical apparatus" (that is, our mental processes). Though psychology differs from physics in that psychology uses its mental processes to study and make inferences about mental processes, physics uses the mental processes to make inferences about the external world. Both sciences often infer what is not directly perceivable. This last Freud describes

---

[10] Freud, *New Introductory Lectures,* 198–200. The two Heine lines translated by Kaufmann in *Freud versus Adler and Jung,* 18.

[11] Sigmund Freud, *The Psychopathology of Everyday Life,* trans. and ed. James Strachey, (Norton, 1960), 134. First published in 1901. The words "forced and tortuous expressions" are Kaufmann's translation in *Freud versus Adler and Jung,* 17.

[12] See above, p. 22.

[13] Sigmund Freud, *Inhibition, Symptom, and Fear,* 231–33, in *Beyond the Pleasure Principle and Other Writings,* trans. John Reddick, (Penguin, 2003). First published in 1926. In this translation, Reddick explains his use of "fear" in the title of the work, as opposed to the *Standard Edition's* "anxiety." Fear, says Reddick, is the typical German meaning of *Angst.* When Freud means anxiety, he uses either *Ängstlichkeit* or *Ängstneurose* for anxiety neurosis, and Reddick translates the words as "anxiety." Note 3, 264–65. References to the *Standard Edition* are to *The Standard Edition of The Complete Psychological Works of Sigmund Freud,* trans. and ed. James Strachey, 24 vols. (Hogarth Press, 1953–74).

as unrecognizable or invisible—"*unerkennbar*" in quotation marks in Freud's German. The *Standard Edition* translation, and, unfortunately, many of the later Penguin translations, is "unknowable," but Freud's meaning is not Kantian.[14] He means that gravity and molecules are no more directly perceivable than the processes of our subconscious minds. We infer them, an apt and accurate analogy of psychology to physics.[15]

In addition, Freud points out, our awareness through sense perception of the physical world, provides "insights into connections and dependencies that are present in the external world," reproduced in our minds as thought and knowledge that enable us not just to understand external reality, but also to predict and change it. And because the procedures of the science of psychology are similar to those of physics, says Freud, our awareness through introspection of the mental world provides similar insight about "connections and dependencies" of the conscious and subconscious mind, to understand both and to anticipate and change, that is, to treat through introspective analysis, our mental functioning.[16]

The thought and knowledge that accumulates in our minds, whether of physics or psychology, consist of concepts and principles, and the process of forming and applying concepts and principles is conceptualization. To put this in Ayn Rand's terms, science is the mental process of identifying the nature of entities and their attributes—physical or mental—including in particular the motions of the entities caused by those attributes. This last is Aristotelian causality based on formal causation, or as Rand puts it, causality is "identity in action."

---

[14] Sigmund Freud, *An Outline of Psychoanalysis*, 14, trans. Helena Ragg-Kirby, in *The Penguin Freud Reader* (Penguin, 2006), ed. Adam Phillips. First published posthumously in 1940. The *Outline* is a concise, though incomplete, summary of psychoanalysis. In the *Oxford-Duden German Dictionary, erkennbar* means recognizable or visible, the negative would be the opposite. The word Kant uses is *unbekannt*, which means unknown. Sometimes, he writes *gänzlich unbekannt*, meaning completely or entirely unknown.

[15] Freud, *Outline*, 13, 50–51. Freud does also say, "The real ['*Das Reale*,' or reality] will always remain '*unerkennbar*,'" but the context is a reference to the physical sciences that must deduce or infer, as the sciences of consciousness must, *the real* that is invisible to sense-perception. Freud, like most laypersons and many non-philosophers, in all probability cannot believe philosophers like Kant really mean what they seem to be saying, namely that true reality is not ever knowable. Quotation marks around *Das Reale* and, again, *unerkennbar* are Freud's.

[16] Freud, *Outline*, 50–51.

Science thus identifies entities and their attributes and causes and effects (connections and dependencies) by constructing and defining concepts that identify the nature and essence of those entities and attributes. To continue the epigraph that heads this chapter, Freud further explains science and scientific method, especially as it applies to psychology and in partial answer to the religious dogmatism of his critics:

> [The scientist] requires a lot of patience, readiness for all possibilities, renounces early convictions lest under their compulsion one should overlook new and unexpected factors; and in the end all this exertion proves worthwhile, the scattered finds do fit together, one gains an insight into an altogether new piece of psychic processes, is done with the task and ready for the next one. Only the help that [controlled] experiments provide for research has to be dispensed with in analysis.[17]

Again this is conceptualization. To add one more point about Freud the scientist, in the words of Walter Kaufmann, Freud often considers "objections and alternatives," many of which we still hear today, but were addressed and answered by him in his time, the accusation of "pansexualism" being just one example.[18] As Kaufmann writes, considering objections and alternatives is the "essence of critical thinking."[19]

---

[17] Freud, *New Introductory* Lectures, 215. Translation by Walter Kaufmann in *Freud versus Adler and Jung*, 79–80. Kaufmann describes this quotation as "the most beautiful formulation of Freud's Goethean and deeply anti-Kantian conception of science."

[18] Preface to the Fourth Edition, 1920, in Sigmund Freud, *Three Essays on the Theory of Sexuality*, trans. and ed. James Strachey (Basic Books, 1962), xxx. First edition published in 1905.

[19] Kaufmann, *Freud versus Adler and Jung*, 80. I would say that the essence of critical thinking is the mental, volitional effort to form and apply concepts and principles that correctly recognize or identify, that is, do not contradict, the facts of reality. It is the ability to perceive reality through unclouded or unstained lenses, or more precisely, to process reality into a mental filing system that captures the real world rather than obscuring it. Considering objections and alternatives is one of the ways we test and try our ideas against reality to ensure there are no contradictions. Kaufmann's original statement on objections and alternatives, an entire sentence italicized in his discussion of Nietzsche, is in *Nietzsche, Heidegger, and Buber*, 129. The sentence ends, also in italics, "while dogmatism ignores both."

Freud did not have a theory of universals, as many working in the special sciences similarly do not have, but he accepted the best of Aristotle to guide his research in psychology, focused carefully on reality to help his patients also come to see, and live in, reality. Freud's view of universals may be described as the typical layperson's—non-philosopher's—understanding of universals and of how we acquire knowledge of concrete reality.

FREUD IN ENGLISH TRANSLATION

Reading Sigmund Freud in English is a challenge. One reason is his approach to the science of psychology and a second more important reason is the English translations.

Freud was trained as a medical doctor specializing in neurophysiology. As a young man, he worked six years in a physiology lab and three more in a hospital where he focused on cerebral anatomy. In 1886, at age 30, he opened his medical practice and soon began treating patients for psychological problems. He did not close the practice until shortly before his death in 1939.[20]

In 1895, Freud drafted a "Project for a Scientific Psychology," premised entirely on neurophysiology, but it was never completed or published in his lifetime, likely because his treatment of patients convinced him of the need to view psychology as a mental science.[21] While Freud in his mature, later life cannot, for the most part, be described either as a materialist or a reductionist, some neurophysiological jargon permeates his work, such as charge and discharge of energy, excitation, and innervation. As a result, a sense of his earlier materialism and reductionism lingers. Although Freud was a determinist his entire life, a reader who is an advocate of free will can overcome this obstacle by focusing on his conceptualizations of psychology and his methods of therapy.

The worst challenge in reading and understanding Freud is the English translations, especially those of the *Standard Edition*, which have been criticized for their misleading translations that have misinformed many English-speaking readers. For this reason, a discussion of the translations at this point is warranted. Although I

---

[20] Peter Gay, *Freud: A Life for Our Time* (Norton, 1988), 649.

[21] Sigmund Freud, "Project for a Scientific Psychology," in the *Standard Edition: Pre-Psychoanalytic Publications and Unpublished Drafts*, trans. James Strachey, vol. 1, 1866–1899 (1954; repr. Vintage Books, 2001), 295–397.

have not studied the German language, I have frequently consulted several online and printed German-English dictionaries, and have also looked at this website, freud2lacan.com, where many of Freud's works are posted as bilingual texts, English on one side of the page, German on the other.

My references for criticism of the translations are four native German speakers: philosopher Walter Kaufmann,[22] Austrian-born psychologist Bruno Bettelheim,[23] Freud biographer Peter Gay,[24] and Berlin-born American social worker H. Frank Brull.[25] The four overlap in their critiques, though Bettelheim's is the most extensive with background information about the literary influences on Freud, such as the classical myths of Oedipus, Psyche, and Eros, as well as Freud's human influences, especially Goethe.

I will focus on mistranslations and neologisms of the *Standard Edition*, but let me first give Kaufmann credit for setting the tone by asserting that the English translations of philosopher Immanuel Kant *improve* Kant's writing style, while the translations of Freud make his *worse*. He also says that Freud's "style is so colorful that the best translations often cannot be more adequate than black-and-white photographs of great paintings."[26] Kaufmann goes on to cite Peter Gay who said the translations make Freud more prolix (diffuse and verbose) and genteel (gentrified, scholarly, academic) than he really is.[27] The word in my mind is sterile, and the translations have loaded Freud's works with "diffuse and verbose" jargon that many of his English and American followers for years unfortunately continued to use.[28]

---

[22] Kaufmann, *Freud versus Adler and Jung*, 20–27.

[23] Bruno Bettelheim, *Freud and Man's Soul* (Vintage Books, 1983). All of this 112-page book is about understanding Freud correctly, discussing many mistranslations and other problems with Freud in English.

[24] Gay, *Freud*, 465–66.

[25] H. Frank Brull, "A Reconsideration of Some Translations of Sigmund Freud," *Psychotherapy: Theory, Research & Practice* 12, no. 3 (1975), 273–79, https://doi.org/10.1037/h0086443.

[26] Kaufmann, *Freud versus Adler and Jung*, 22.

[27] Kaufmann, 23. Peter Gay, *Freud, Jews, and Other Germans* (Oxford University Press, 1978), 41n20.

[28] Kaufmann, I should further point out, makes it clear that Freud was neither dogmatic nor intolerant. The smears against him—in addition to the suspicion, laughter, outrage, and disgust thrown at psychoanalysis by the local Viennese, many of whom were antisemitic and had not read Freud—arise from the

Bettelheim says the translations make Freud impersonal and abstract with Latinized and Greekized words, and sometimes translate him in the passive voice when Freud uses the active. Freud is even made to sound behavioristic, apparently in order to make him more "scientific," which in philosophic terms means more positivist, which Freud was not.[29]

Bettelheim became aware of the problems with the English translations when he was at the University of Chicago's Orthogenic School from 1944–73. His staff and students, as non-German speakers, read Freud in English, which made them view him and psychoanalysis distantly—as "a way of looking at others," especially patients, in a cold, reserved manner, that is, "from a safe distance."[30]

The most egregious mistranslation is of the German word *Trieb*, translated throughout as "instinct." The real meaning is *drive*, the urge to act when a need is not being satisfied. The hunger drive is offered by Freud as analogous to his life and death drives. He rarely uses the German equivalent for instinct (*Instinkt*) and only to refer to the alleged automatic knowledge that guides animal behavior.[31]

The use of "instinct" makes Freud look unscientific and like an advocate of innate knowledge. Freud, in fact, uses the term "drive" as a cause that makes "physical demands upon the psyche," such demands being changeable and competitive with one another.[32] They are not instincts. Bettelheim insists that the title of the 1915 *Instincts and Their Vicissitudes*, one of Freud's later works, should be translated as "Drives and Their Mutability."[33]

---

resentments and complaints of Freud's former devotees, Alfred Adler and Carl Jung. Kaufmann, *Freud versus Adler and Jung*, 195–204, 325–50. Adler, in particular, referred to psychoanalysis as "filth and fecal matter." Kaufmann, 199. Gay, *Freud*, 410, acknowledges that Freud is a "determinist, [but] not a fatalist," a reference to the misunderstanding that we are allegedly completely controlled by the unconscious, which ignores Freud's frequent assertions that the goal of therapy is to make the unconscious conscious.

29 Bettelheim, *Freud and Man's Soul*, 5–8, 108. To borrow F. A. Hayek's word *scientism*, the English translations of the *Standard Edition* seemed intent on making Freud sound *scientistic*.

30 Bettelheim, 6.

31 Today, the concept "instinct" should be dispensed with as invalid. Biologists have made great progress in explaining animal behavior, whether it is how to hunt (they learn from their parents, siblings, or by trial-and-error), to migrate to warmer weather (birds use information from the sun, stars, and magnetic field), or to find their way home (cats use the magnetic field and scent).

32 Freud, *Outline*, 4.

33 Bettelheim, *Freud and Man's Soul*, 105.

The most egregious neologism—this time "egregious" is Peter Gay's word—is the Greek import *cathexis*. Freud wrote using common German words, *Besetzung* being one of them with many connotations. The meaning of the word as Freud used it is occupation, as in military occupation. It is sometimes translated into other languages as load or investment (of emotion). Freud suggested the word "interest," and in the 1926 *Encyclopedia Britannica* summary of psychoanalysis, written by Freud, the words "affective charge" are used.[34] *Besetzung* is a term of value and emotion and the "occupation" is a quantity—high or low, strong or weak—of emotional energy.

Thus, I might say I am highly "cathected" in the family dog, or rather, "occupied" with a lot of value and emotion in him. My brain, however, still freezes when I come upon cathexis or any of its word forms (hypercathexis, anticathexis, etc.) in the translations of the *Standard Edition*. I find myself having to pause to review the context of the word before continuing. Another neologism, *parapraxis*, can be dispensed with quickly. *Fehlleistungen* means faulty action or achievement, or in common English parlance Freudian slip—of the tongue, pen, memory. Why the Greek?[35]

Other mistranslations: id, ego, and superego in German are *das Es, das Ich*, and *das Über-Ich*.[36] They should be what they are, and are so translated in the French and Spanish editions: *the it, the I*, and the *upper-I*. The "it," a word used by both Friedrich Nietzsche and a contemporary physician of Freud's, was chosen to describe the unconscious because Freud's patients would often say something like, "*It* got me again" or "*It* makes life unbearable." "*The I*," in German can mean

---

[34] Sigmund Freud, "Sigmund Freud on Psychoanalysis," *Encyclopedia Britannica*, November 6, 2014, britannica.com. First published in English in 1926.

[35] One other Greekized and bothersome neologism is *anaclisis*, which translator John Reddick says is a "startling misunderstanding of the German," defined in the *Standard Edition* as attachment and today as psychological dependence. Reddick, in his translation of Freud, *On the Introduction of Narcissism*, 243n27, in *Beyond the Pleasure Principle and Other Writings*, points out that in Freud's usage, *Anlehnung* meant imitative or modeled on, meaning, for example, that children unconsciously imitate or model their parents' values and behavior. The narcissism monograph was first published in 1914.

[36] Bettelheim, *Freud and Man's Soul*, chap. 8. Translator Helena Ragg-Kirby retains the German, *das Es*, etc., in her translation of the *Outline of Psychoanalysis*. I found that that usage made Freud more approachable than the Latinizations. Generally, I think the Penguin Books series of retranslations from the early 2000s are more readable than the translations of the *Standard Edition*. Unfortunately, they lack indexes, which I consider essential aids when reading scholarly books.

the self or personality; "self" is the word I substitute when reading the English translations. *Über* can mean above or over, but I like Bettelheim's "*upper-I*," because it emphasizes that our conscience is a part of our minds (implying that we put it there), not some compartment separate from the self or mind. The Latinizations make Freud and our understanding of the mind sterile. Cold technical jargon, as Bettelheim says.[37]

One final mistranslation in this section; more will be addressed as they come up in the remainder of the chapter. The word Freud uses for the subject of psychology is *Seele*, which means soul, and *Seele*, Freud points out, is the German translation of the Greek word for soul, *psychē*, which Aristotle used. Thus, psychology is the science of the soul, not mind or consciousness, though it can include both, and soul is neither immortal nor related to religion. For Freud, according to Bettelheim, soul means "that which is most valuable in man while he is alive." It is not just the seat of reason, but also, and perhaps more importantly, the seat of emotion. It defines who we are and what motivates us.[38]

To conclude this section I do have something positive to say about the *Standard Edition*. James Strachey was editor and translator, and the work was a major accomplishment—twenty-four volumes published between 1953 and 1974, every volume with an excellent index and an abundance of footnotes and bracketed comments that explain, for example, someone from Freud's era whom we would not today recognize. It is, or was, the most significant scholarly work in English on Freud. *The Revised Standard Edition of the Complete Psychological Works of Sigmund Freud* has recently been published, in June 2024.[39]

---

[37] Bettelheim, *Freud and Man's Soul*, 53. "Turgid and often obfuscatory style" is how translator John Reddick described the *Standard Edition*. Translator's preface to Sigmund Freud, *Beyond the Pleasure Principle and Other Writings*, xxxiii.

[38] Bettelheim, *Freud and Man's Soul*, 77. Cf. Branden, *Psychology of Self-Esteem*, 132. Defense and repression are two translated words I find acceptable in English, though Bettelheim, 91–94, provides helpful information about the meanings of the German words. *Abwehr*, defense, is the broader one and means a repulsive action to fend off or parry anxiety to protect the self, whereas *Verdrängung*, repression, means to displace or drive the anxious event out of consciousness into the unconscious.

[39] I have only seen brief excerpts of this edition. It seems to be an improvement, a 30-year "mammoth task," said editor Mark Solms. The style is still similar to that of James Strachey, but the word "instinct" has been replaced by "drive."

As can be seen in this discussion of rendering Freud's German into readable English, translation is highly epistemological, a "reconceptualization" as one translator describes his craft.[40] Translators must convey the meaning of words that represent concepts, the ideas that stand behind the words. As Ayn Rand points out, there is a range of options in how concepts can be formed, which means other languages must be translated into one's own by using, in Rand's words, "descriptive phrases" to convey "subtle shades of meaning."[41] Neologisms, generally, are not a good approach to translation and mistranslations result from a lack of sound epistemology.

THE BASIC SCIENCE OF PSYCHOANALYSIS

The fundamental principle of the theory of psychoanalysis is the influence on our adult psychologies of healthy and unhealthy responses to childhood experiences, stored in the unconscious. Freud is not the first to suggest this notion, but he is the first to provide highly detailed discussions of it. By rescuing the concept of neuroses from the merely physical view that they are nervous diseases, he turned psychology into a mental science.[42] Let me now summarize Freud's key ideas—an essentialization, not an exhaustive presentation.[43]

Fundamental Concepts

Freud says that human beings possess needs expressed as drives, or urges to act, to satisfy those needs. The process requires mental activity to identify what will satisfy the need and decide upon action to attain it.

Other Strachey terms have been retained. The index was thoroughly reworked, as were the footnotes and editorial introductions. The translations were reviewed line by line. And previously untranslated material has been added. Sophia Stewart, "Freud's Writings Get an Update—30 Years in the Making," *Publisher's Weekly*, August 1, 2024. Mark Solms, "Notes on the Revised Standard Edition," *Psychoanalytic Review* 100, no. 1 (February 2013), 201–10, https://doi.org/10.1521/prev.2013.100.1.201. For an earlier comment, see Douglas Kirsner, "Fresh Freud: No Longer Lost in Translation," *Psychoanalytic Psychology* 24, no. 4 (2007), 658–66, https://doi.org/10.1037/0736-9735.24.4.658.

[40] Mark Mellor, "Elusive Objects of Translation," *Cadenza Academic Translations*, cadenza-academictranslations.com.

[41] Rand, *Objectivist Epistemology*, 72.

[42] Cf. Lancelot Law Whyte, *The Unconscious before Freud: A History of the Evolution of Human Awareness* (Basic Books, 1960).

[43] Which means some debatable or controversial issues are omitted in this summary, though I find those issues also interesting to read and think about.

The causal impetus from these mostly unconscious biological drives is what Freud calls the "pleasure principle," though a better translation of *lust*, according to Brull, would be joy or desire.[44]

For example, when dying of thirst on a hot summer morning, I often feel a strong desire for a cold glass of water and to experience joy in its consumption. The "reality principle" of our reasoning mind, however, can interfere with, detour, or adapt the drive because of what is required for daily life in the world.

The I may remind us, in other words, as it often does me, that drinking the cold glass of water was not a good idea and that sips of water with food, to lessen the coldness, would have produced a better outcome. The *upper I*, our memories of parental influence (plus other influences), may also have something to say about this, reminding us not to drink too much water on an empty stomach in the morning, unaccompanied by food.

This in a highly simplified nutshell is the essence of Freud's theory of normal psychology. Neurosis then arises from deficiencies in meeting our innate psychological needs.

Hunger and thirst, of course, are not the drives that Freud focuses on, though he does use them as analogies.[45] His main drive is the *libido*, a Latin word adopted by him that means wish, desire, or love. In his younger years, he used it to mean the sex urge, but he broadened it later to mean emotional energy that originates in our sexuality, expressing our sense of self. When connected to the *I-drive*, it emphasizes and expresses self-assertion.[46] Libido is the life force or love drive that motivates us.

Sense of self can be interpreted as self-esteem, which Freud was not unaware of as a psychological need, and he did observe its deficiency in many of his patients. Freud, however, did not emphasize self-esteem as a need or drive, though in several places he quotes patients expressing

---

[44] Brull, "Reconsideration of Some Translations," 277.

[45] See especially Freud's discussion of the evolution of psychoanalytic theory that began with what he calls the "popular" distinction between the drives of hunger and love. Sigmund Freud, *Beyond the Pleasure Principle*, 90–92, in *Beyond the Pleasure Principle and Other Writings*. First published in 1920.

[46] Self-assertion is Brull's preferred translation of *Selbstbehauptung*, as opposed to the *Standard Edition* rendition as "self-preservation." Self-assertion is also the only entry in the *Oxford-Duden* dictionary for *Selbstbehauptung*, with claim or assertion as the meaning of *Behauptung*.

precisely what psychologist Edith Packer calls core evaluations.[47] "I always had a low opinion of myself," says one patient.[48] And "I can't accomplish anything, I can't succeed in anything," says another. Loss of love in childhood, says Freud, can lead to "a legacy of diminished self-feeling amounting to a narcissistic scar," the German word for self-feeling being *Selbstgefühls,* which has been translated as self-esteem, though the former is probably more correct.[49]

Self-feeling, rather than self-esteem, is more in line with Freud's concept of narcissism, a self-absorption of children, and adults who are physically or mentally ill. That is, narcissists lack maturity (in the case of children) or health (in adults) to allow them to focus on and love—that is, to invest one's libido in—other people or even material goods to enhance their lives. Being in love, says Freud, increases our self-feeling by reducing narcissism. Suffering neurosis, however, reduces self-feeling, which leads to various symptoms, including the sense of inferiority that most neurotics experience.[50]

Narcissism for Freud is inherent in all of us from birth, as part of the libido drive, that most do not and probably cannot fully overcome. Sublimation means elevating or rising above the lower biological drives to something higher, for example, pursuing de-sexualized artistic or intellectual careers. Such paths, however, because of Freud's determinism, are not open to everyone.[51]

In my words, I would say that the libido drive is the pursuit of values in oneself and in the pleasure of the company of others, which would include the pursuit of romantic values. The pursuit of mistaken or unhealthy values then is what leads to neurosis. In childhood, this pursuit of values includes the exploration of our sexuality.

According to Freud, sexuality develops in childhood in three stages of sensuousness: the oral, sucking of the breast and thumbs; the sadistic-anal, of learning to control defecation (and maybe even playing with

---

[47] Packer, *Lectures on Psychology,* chap. 1.

[48] Sigmund Freud, "Psychotherapy of Hysteria," 278, in Josef Breuer and Sigmund Freud, *Studies on Hysteria,* trans. and ed. James Strachey (Basic Books, 1955). First published in 1895.

[49] Freud, *Beyond,* 58–59, in *Beyond the Pleasure Principle and Other Writings.*

[50] Freud, *Narcissism,* 11–20, in *Beyond the Pleasure Principle and Other Writings.* Feelings of inferiority that Freud writes about are the not the same as Alfred Adler's inferiority complex.

[51] Freud, *Narcissism,* 23–24, in *Beyond the Pleasure Principle and Other Writings.*

the feces) coinciding with teething and therefore "sadistic" biting;[52] and the phallic (genital), with all the questions about where babies come from and the differences between the sexes that follow, and of course, the pleasure of touching the genitals.

That sexuality begins before puberty was radical for Freud to assert in his time. Sexuality, as he points out, is a broader term than the sex or sexual relations of adults. His *Three Essays on the Theory of Sexuality* is a daring and non-judgmental (that is, de-moralized) masterpiece for his era, covering all forms of sexuality, which include all the known deviations from penile-vaginal intercourse. A fixation on the foot or a piece of clothing—fetishes, in other words—or inversions (for example, voyeurism at the same or opposite sex), Freud points out, develop early and cause neurotic problems in adulthood.

Freud focused so much on sex because his Victorian culture produced problems in his patients that derived from their early childhood experiences, and especially from the often negative reactions of their parents. Women, for example, may not have been taught or have known anything about sex,[53] and young men, in addition to having suffered threats as children of having their penises cut off, were considerably bothered by coitus interruptus, the withdrawal method of birth control and frequent cause of impotence.

When the libido, called by Freud in later years the life or *Eros* drive, misfires, the death or destructiveness drive leads us to a "flight into illness" or "withdrawal from reality."[54] The misfiring produces "excitation" (one of Freud's neurophysiological terms) experienced during disturbing or frightening events, often in childhood, that

[52] I would put the "terrible twos" in this stage, which is the beginning of self-assertion, but Freud does not explicitly mention it.

[53] Josef Breuer in "Fräulein Anna O." 21, Breuer and Freud, *Studies on Hysteria*, describes his treatment from 1880–82 of a twenty-one-year-old woman whose "sexuality," he wrote, "was astonishingly undeveloped." The woman was Bertha Pappenheim, who later became a social worker and feminist advocate for Jewish women. In print, she was referred to as Anna O. Freud credits Breuer as the person who convinced him of the value of talk therapy.

[54] Sigmund Freud, *Five Lectures on Psycho-Analysis*, trans. James Strachey (Norton, 1961), 54. First published in 1910. The lectures were delivered extemporaneously in 1909 at Clark University in Worcester, Massachusetts, Freud's only visit to the United States.

were not dealt with appropriately at the time. Thus, the emotions of multiple events are condensed and displaced into a symptom, while the ideation of the events is repressed.[55]

Neurotic anxiety, a fear that has no apparent object or cause, is often experienced in response to the "excitations" of disturbing events. Ideations associated with the events are pushed out of conscious awareness, not recurring sometimes until adulthood and then only as neurotic symptoms. Repression is the fundamental defense that *the I* uses to protect itself, but there are other defenses, such as reaction formation, projection, and regression, as well as displacement and sublimation.

Most of these concepts of defense—reaction formation the exception—predate Freud, but he gave them new and more developed meanings. The Oedipal complex is original to him and is Freud's metaphor for childhood injury that results from parental reactions to the child's overinvestment of libido in one parent, often, though not always, the mother, and underinvestment in the other, the father.[56]

Dreams

Freud did not begin his career analyzing dreams. At first, he worked with hysteria patients, the word "hysteria," or "conversion hysteria," in his day essentially meaning psychological problems that have been converted into psychosomatic symptoms. In the course of working with these patients, he found that dreams were not just relevant, but their analysis was an aid to patients in uncovering and changing early childhood errors. Thus, the subject that first made him famous, dreams and dream analysis. Let me summarize his accomplishments on dreams. It represents an outstanding epistemological analysis, classification, and conceptualization of the subject.

---

[55] *Vorstellung* is the word for "idea," a philosophical term used, for example, to translate John Locke's English word "idea" into German. Sometimes, it is translated as presentation or representation. Freud frequently uses *Vorstellung* to mean the idea or ideation behind an event, especially behind the emotion of an event.

[56] See Bettelheim, *Freud and Man's Soul*, chap. 5, for a thorough discussion of the Oedipal Complex and its background in the classical myth.

As paraphrased by Freud, Aristotle said dreams are "mental activity of the sleeper in so far as he is asleep."[57] They are not supernatural messages from the gods. Freud's study of dreams said they are harmless hallucinations when asleep, unconscious urges expressed as a wish or urges from our "day's residues" (waking life) expressed as a wish, or some combination.[58] He did not say that dreams are all about sex, as critics have alleged, and symbolism, sexual or otherwise, must be related to the dreamer's private meanings, that is, the symbols are not universal.

By analyzing a large number of dreams of both normal people and people with psychological problems, Freud provided a valuable and significant contribution to psychology: the distinction between primary and secondary mental processes of consciousness, and a better understanding of the unhealthy processes.

For dreams in particular, he gave detailed descriptions of their nature, including a terminology to use for analysis and a three-part classification of them. "Manifest content" is the actual events of the dream that we experience, weird and distorted as they may be. "Latent content" is the underlying thoughts that evoked the dream. And "dreamwork" is the mental process that transforms the latent into the manifest. The work of dream analysis is to trace this transformation and thereby identify the meaning of the dream.[59]

Manifest dreams, Freud identified, are condensed, versus the actual events of a memory, and exhibit a great deal of displacement, that is, the shifting or substitution of minor elements from reality

---

[57] Sigmund Freud, *The Interpretation of Dreams*, trans. and ed. James Strachey (1955; repr. Basic Books, 2010), 36–37. First published in 1899 with a publication date on the title page of 1900, to bring in the new century. Though Aristotle is mentioned in the first edition of the book, this definition attributed to him does not appear until the fourth edition in 1914. Freud was not just a prolific writer, but a prolific editor of his previously published works, updating and adding material. *The Interpretation of Dreams* went through eight editions in his lifetime. Shorter discussions of dreams can be found in *On Dreams*, published in 1901; in part 2 (eleven chapters) of the *Introductory Lectures on Psycho-Analysis*, 1916–17; and in chapter 5 of *An Outline of Psychoanalysis*. Short comments on dreams are also made throughout his other works. See Aristotle, *On Dreams*, 462a15, in McKeon, *Basic Works*, 625.

[58] Freud, *Interpretation*, 79, 114–17. Freud, *Outline*, 17.

[59] Sigmund Freud, *On Dreams*, trans. and ed. James Strachey, (Norton, 1952; reprint, 1989), 16. First published in German in 1901.

to major ones in the dream or vice versa, or some mixture. Often, a single dream can have several elements, including contraries or contradictions, side by side one another.

The simplest and most fundamental dream is the child's direct wish fulfillment. For example, a little girl who was ill during the day and forbidden to eat strawberries experienced an elaborate dream of eating "stwawbewwies and omblet" during the night. And a little boy whose hike to a nearby mountain was cut short for lack of time dreamed of his conquering the mountain.[60]

Of course, adults can experience direct wish fulfillments, such as having to go to the bathroom in the middle of the night and dreaming of sitting beside a waterfall, or of being hungry and dreaming of a gourmet meal, or, yes, of dreaming about sex. Criminals also dream about escaping through their barred windows. Most of our dreams, however, are distorted. Why?

In addition to the direct wish fulfillment, Freud classifies dreams as either anxiety or punishment dreams. In my case, if I may cite a recurring dream, I retired from teaching in 2015 but find myself since dreaming about being late to class and/or unprepared, which never happened in thirty-six years of teaching. And in some versions, the building I am belatedly trying to get into to teach a class is the junior high school I attended in my hometown. (Anxiety dreams can be experienced as nightmares, though I have not experienced these dreams as such.).

The underlying dream thoughts: I do have an anxiety about being late to any appointment and prefer to arrive five to ten minutes before the appointed hour. How is this a wish? Freud would say that when asleep my subconscious brings up the repressed anxiety and emotions of embarrassment and humiliation about being late, then distorts the wish of being on time, turning my latent dream thought into a distorted manifest dream, perhaps as a caution to be extra careful about appointments.

And the junior high school building? I think there is an element of direct wish fulfillment here, as I do have a fondness for the building. Both of my older brothers attended it when it was a senior high school and I attended many an enjoyable performance in the school's auditorium, plus basketball games in the gymnasium and high school football

[60] Freud, *On Dreams*, 20–21.

games in the nearby stadium. But no, I never had a desire to teach in that building. One might say, "all school buildings look alike," hence the displacement.

Punishment dreams often dredge up unpleasant events from the past that the dreamer experienced as embarrassing or humiliating, sometimes resulting in a sense of guilt. A well-known author in his younger years was a tailor, which he preferred not to think about, but repeatedly dreamed about those earlier years. Similarly, Freud admits that he was not particularly competent when working as a young man at the Chemical Institute, yet in his successful older years continually dreamed about those humiliating days.[61] How is this wish fulfillment?

Punishment dreams, says Freud, are not wishes of subconscious drives or memories, but of "a special critical and prohibiting agency" in the mind, *the above-I*.[62] Punishment dreams are a wish of punishment arising in our conscience for the sense of guilt from earlier days.

At this point, Freud discusses the issue of traumatic dreams, especially those of war trauma or, in modern terms, post-traumatic stress disorder (PTSD). Trauma is experienced as the emotion of fright, which is terror, an extreme fear, deriving from unpreparedness for and surprise at the severity of the threatening event. Freud concedes that it is difficult to call these dreams wish fulfillments and suggests that the dreamwork fails to transform the memories into a wish fulfillment.[63]

Let me object, though, and suggest that, as Freud acknowledges, all psychological processes exist on a continuum from normal to abnormal, thus the punishment and traumatic dreams may be variations on the anxiety dreams. We all differ in how we react to negative events in our lives, depending on the level of self-esteem we each have, other inner resources, and the severity of the trauma. And there are war veterans who do not suffer PTSD.[64]

Sleepers who have traumatic dreams are often awakened feeling acute anxiety, because the dream has been an exact replica of the traumatic event. The dream and anxiety indicate the amount of fright

---

[61] *Interpretation*, 481.

[62] Freud, *New Introductory Lectures*, 34.

[63] Freud, *Beyond*, 50–52, 71–73, in *Beyond the Pleasure Principle and Other Writings*. Freud, *New Introductory Lectures*, 35–37.

[64] Freud, in his introduction to *Psycho-Analysis and the War Neuroses*, ed. Ernest Jones (International Psycho-Analytical Press, 1921), 4, suggests that war trauma is on the extreme end of the continuum from elementary to advanced neuroses. "Every neurosis," he says, is a "reaction to a trauma," gutenberg.org.

experienced during the initial event, along with the lack of preparation for such a violent event. Afterwards, an emotion of guilt that says, "why me?" can easily arise—that is, a guilt for having survived the event when one's comrades did not. Punishment dreams? Sounds like it.

Primary-Secondary Mental Processes

Freud's work on dreams, answering the critics in his time who regarded them as meaningless "froth," elaborated Aristotle's identification of dreams as mental activity while asleep. His work led him to make a significant distinction between the actions of our unconscious mind and the level of conscious reason. He called the former "primary process," because it arose in the evolutionary process before our human conscious mind, which he called "secondary process." The higher animals exhibit a primary-process consciousness, and a modicum of choice, but not a true, self-aware volitional consciousness, though Freud did not put it in those words.

Primary process seems random, meaning no predictable pattern, but Freud demonstrated that dreams are not random. He recognized that the source of psychological symptoms, both neurotic and psychotic, this last called by many thinkers "waking dreams," can be found in our unconscious primary-process minds.[65] Which is why Freud often wrote that the goal of psychotherapy is to "make the unconscious conscious." Later psychiatrists, Eilhard von Domarus and Silvano Arieti, based on their work with schizophrenics, offered additional explanations of how our subconscious mental processes do have a logic to them, albeit a lesser logic than what our conscious minds are capable of generating.

Psychiatrist von Domarus hypothesized that the subconscious, at least in part, citing what is now called the Von Domarus principle, is guided by the formal fallacy in logic of the undistributed middle term.[66] For example, in the syllogism—all dogs are four-legged animals, all cats are four-legged animals, therefore all dogs are cats—the middle term is the one that appears in each premise but not in the conclusion. Four-legged animals is the middle term that is not distributed, which

---

[65] Freud, *Interpretation*, 115.

[66] Eilhard von Domarus, "The Specific Laws of Thought in Schizophrenia," ed. J. S. Kasanin, *Language and Thought in Schizophrenia* (University of California Press, 1944), 104–14.

means that because four-leggedness is in the (grammatical) predicate of both premises, neither account for all four-legged animals. That dogs and cats have four legs does not mean they are otherwise the same.

The fallacy is one of overgeneralization—assuming that because two things have one attribute in common, they are overall the same. Young children, until they learn better, make the dogs-and-cats type of mistake. Members of indigenous tribes (or other adults who are being careless) also make the mistake. And schizophrenics have drawn conclusions like this: the Virgin Mary is a virgin, I am a virgin, therefore I am the Virgin Mary. The predicate term "virgin" is not distributed. Or, they have said: Switzerland loves freedom, I love freedom, therefore I am Switzerland.

Arieti, agreeing with von Domarus, called the logic of our subconscious primary process "paleologic" and the reasoning of our conscious secondary process "Aristotelian."[67] The former is paleo because it is old or archaic, predating our educated control of thoughts using what Aristotle was first to identify, especially his three laws of logic. Children have not yet learned the skill of competent reasoning and schizophrenics in their psychotic episodes are living "waking dreams," having fled "the unsatisfying reality," and are controlled entirely by their subconscious.[68]

Arieti recognized that the errors of schizophrenics are errors of association and generalized them, in somewhat less technical terms than the fallacy of undistributed middle, as *the identity of predicates*. The schizophrenic, young child, and indigenous person assume that because the predicates in their thoughts are identical, their subjects must overall be the same.[69]

---

[67] Silvano Arieti, *Interpretation of Schizophrenia*, 2nd ed. (Basic Books, 1974), 229–41. First edition published in 1955.

[68] Freud, *Five Lectures*, 54.

[69] A valid Aristotelian syllogism, for example, would be: all dogs are four-legged animals; Fido is a dog; therefore, Fido is a four-legged animal. "Dog" is the middle term and is distributed in the major premise (the first one), because it tells us *all* dogs are four-legged. Aristotelian logic focuses primarily on the classification of subjects, rejecting the identity of predicates. In this syllogism the subject Fido is a member of the category "dogs" and therefore has four-legs. Arieti, *Interpretation of Schizophrenia*, 236, is mistaken when he writes, "In Aristotelian thinking only identical subjects are identified." It is more complicated than that. In the above syllogism, the subject of the conclusion is identified as belonging to the predicate of the major premise.

When we are fully conscious, our subconscious is still operating and can make seemingly random associations, but the associations are not totally random. For example, a friend at dinner complains that the food is not well cooked, so I might say, "By association, that reminds me of last week when I was served almost raw chicken," the predicate being poorly cooked food. And we often experience similar kinds of associations.

What determines the selection of predicate, considering that there exists a nearly unlimited field of possible predicates? "Emotional factors," says Arieti.[70] These emotional factors are especially important in understanding schizophrenics, who frequently come up with bizarre connections not unlike those made in dreams. I previously made passing reference to one's "private meanings" in understanding the meaning of symbolism in dreams. Knowing a person's private meanings by reference to his or her emotions and emotional generalizations, helps explain seemingly arbitrary associations.[71]

What these associations of the subconscious mind are not are the associations—let me now call them connections—of concept formation, the secondary process of the reasoning mind.

Two Kinds of Association

We now are talking about two different kinds of association. The conscious mind is able to both connect and keep separate two entities with the same characteristic. The dog has four legs and the cat has four legs, but a dog and a cat are not the same. When we form concepts we look for fundamental or essential similarities.

Ayn Rand gives us the rule of fundamentality to enable us to dismiss four-leggedness as the most essential characteristic of dogs. Or, opposable thumbs, for another example, as the essential distinguishing characteristic of human beings.[72] The rule also allows us to dismiss the fact that all humans are capable of building roads, which other

[70] Arieti, *Interpretation of Schizophrenia*, 236.

[71] "Freud's Method for Interpreting Dreams," *Freud Museum London*, freud.org.uk. By association, the words "private meaning" reminds me of Alfred Adler's *private logic*, which, as I wrote in *Independent Judgment*, 65–66n26, means "neurotic reasoning, [though Adler's] followers . . . broadened [the term] to cover thinking patterns of both normal and abnormal psychologies." I suggest on those pages that private logic, as used by Adler's followers, is possibly the source of the notion that Rand's followers call psycho-epistemology.

[72] Rand, *Objectivist Epistemology*, 45–46.

animals cannot do, because "building roads" is not the essence of being human. The rule says that the essential distinguishing characteristic of a concept must *explain and cause* all or most of the other distinguishing characteristics. This gives us the *fundamental* characteristic(s) and only the capacity to reason meets the rule—and explains and causes the ability to build roads.

This process is the operation of Aristotelian logic at its best. The conscious secondary process gives us control, should we choose to exercise it, over the content and processes of our minds. The paleologic of the subconscious mind cannot, and does not, follow the rule of fundamentality. It follows the rule of "identity of predicates."

Having said this, we can to some extent still control our subconscious mind. If we focus and volitionally exercise our conscious Aristotelian logic over time, we can correct some overgeneralization errors of the subconscious.

We can also influence the subconscious in problem solving during the creative process. When working on a problem—say, writing an essay—we usually beforehand must do a certain amount of research and thinking about the issue, building up a substantial context for the problem. To sleep on it, a suggestion some of our teachers may have taught us, now allows our associational subconscious to go to work. Sometimes, not always, we may have a solution or part of a solution the next morning. And, occasionally, when I've seen a particularly distressful movie, I have been able to tell my subconscious at bedtime, "I don't want to dream about this," and often it works.

The significance for Freud of primary process in its dream operations, for purposes of psychotherapy, is that primary process operates in the formation of neurotic symptoms, namely condensation and displacement. What seem to be random associations are condensed and displaced emotions with the ideation, the ideas originally associated with the emotions, repressed. And, identity of predicates and overgeneralization are more specific operations in the formation of these symptoms, which now takes us to the applied science of psychoanalytic therapy.

THE APPLIED SCIENCE OF PSYCHOANALYTIC PSYCHOTHERAPY

Deficiencies in meeting our human requirements for a healthy life, that is, our physical and mental needs, call for treatments to effect a cure for the deficiency or at least a removal of pain. Medical doctors are applied scientists who treat physical ailments. Psychotherapists are mental doctors who treat mental difficulties. Both presuppose the fundamental principles of their respective basic sciences. Both apply those and other principles unique to their fields to promote health in their patients. Both are using concepts of method to identify the steps required to reach their respective goals.

Psychoanalytic therapists take Freud's basic concepts of psychoanalysis, along with specific knowledge of each patient's personality, character, and history, and apply these notions to help the individual patient become happier and more independent.

Making the Unconscious Conscious

The essence of psychoanalytic therapy, according to Freud, is talk therapy with the goal of making the unconscious conscious. It means unearthing or digging deep—he often uses archeology as a metaphor for psychotherapy—in the unconscious to find repressed ideas associated with disturbing events and to encourage the patient to re-experience the original emotions. This "abreaction"—*abreagieren* in the German, which means to work off or get rid of—can be described as an early understanding of de-repression. In later years, Freud devalued abreaction and, in its place, emphasized working with the patient to overcome resistance sufficient to recall and talk about what was painful and repressed.[73]

The treatment method of talk therapy, as I mentioned earlier, was learned from Josef Breuer who encouraged his patient Anna O. to reminisce about early disturbing experiences that might have led to her conversion hysteria (paralyzed arm among other psychosomatic ailments).[74] In the process of this therapy she said jokingly that she was chimney sweeping, but more seriously described the method as the talking cure. In the early years, Breuer and Freud called this the cathartic method, a cleansing or purifying of the soul. For some years, Freud used hypnosis to overcome resistance, gradually abandoning it in favor of the talk method of finding connections to early and forgotten events, thoughts, and emotions.

In the last two sentences of lecture 31 in Freud's *New Introductory Lectures,* Freud writes, "Where *it* was, there should become *I.* [Therapeutic effort] is a cultural achievement somewhat like the draining of

[73] Freud, "Remembering, Repeating, and Working Through," 33–34, and *Inhibition, Symptom, and Fear,* 219, in *Beyond the Pleasure Principle and Other Writings.*

[74] Breuer, "Fräulein Anna O." 30, in Breuer and Freud, *Studies on Hysteria.*

the Zuyder Zee." That is, just as the Dutch reclaimed land from the sea, so also therapists help patients reclaim their conscious souls (*the I*) from the unconscious *it*.[75]

An important concept of therapy that Freud discovered by chance was transference, the idea that patients, as they improve, begin to associate the therapist with an important person from childhood, usually a parent, thus transferring their feelings about this person, both good and bad, to the therapist.[76] Often the patients feel an unconditional love from the therapist that they likely did not experience from their parents.[77] Sometimes, though, anger and hostility may be felt, depending on the type of childhood the patient experienced. And on occasion a patient of the opposite sex may fall in love with the therapist—one young female patient flung her arms around Freud's neck, convinced that they were in love with each other, and Breuer's patient, Anna O, near the end of her treatment supposedly persuaded herself that she was carrying his baby.[78]

Once Freud understood these transference reactions, he identified their value in therapy. When the patient realizes these feelings toward the therapist are displacements, the patient becomes more motivated to overcome resistance to probe more deeply into the unconscious to identify forgotten, repressed thoughts and feelings.

The applied science of psychotherapy as developed and practiced by Freud does not differ much from what nearly all effective psychologists and psychiatrists do today. Freud's so-called free association is a probing of the patient's subconscious in an uncritical, unjudgmental manner to encourage thoughts and emotions to rise to consciousness.

---

[75] Bettelheim's translation. The sentences are on two different pages in *Freud and Man's Soul*, 61, 62. My italics. In Freud the two sentences are consecutive. In the *Standard Edition* of the *New Introductory Lectures*, 100, the first sentence reads, "Where id was, there ego shall be." In *Independent Judgment*, 176, I rephrased that sentence in an updated form using the psychology of Edith Packer: "Where subconscious, mistaken conclusions were, there confident and independent self-assertion shall guide." See Bettelheim's discussion, 61–64, of Goethe's *Faust*, which Freud knew well, including Faust's struggles to reclaim his soul from the devil, the plight of Faust being analogous to what many of Freud's patient's felt, and most likely, what many patients today feel.

[76] Freud, *Outline*, 29–32, in the *Penguin Freud Reader*. Countertransference, going in the other direction, means that feelings from the therapist's psychology and background are transferred to the patient.

[77] Which Carl Rogers came to call "unconditional positive regard," *On Becoming a Person: A Therapist's View of Psychotherapy* (Houghton Mifflin, 1961), 61–62.

[78] Reported years later in a letter by Freud to Stefan Zweig. Gay, *Freud*, 67.

Freud's technique, as becomes clear in his many discussions of therapy, is not the caricature we sometimes see today with the therapist saying repeatedly, "uh-huh, uh-huh," and not much else. In fact, Freud often explores a patient's line of thought and offers hypotheses about earlier unconscious ideation. His goal, after all, is to help patients gain or regain their independence—Freud's word.[79]

Kaufmann, in the first volume of his trilogy, discusses Goethe's concept of autonomy as standing alone against the crowd. He also discusses how this notion influenced Freud, expressed notably in Freud's autobiography where he acknowledges his own development in achieving a "certain degree of independence of judgment" in the face of the University of Vienna's antisemitism.[80] Borrowing Henrik Ibsen's words, Freud called the opposition to himself the "compact majority." Goethean autonomy or independence is the goal of psychoanalytic therapy.[81]

The Freudian couch? Initially a gift from a patient in 1890; the carpet covering added by Freud "gave his patients a non-medical bed to lie on" and "the sense of being sheltered from one's daily cares."[82] In addition, many patients were female and, in his day, understandably uncomfortable talking about sex while looking at a man, nor did Freud want his patients attempting to read his facial expressions. Freud did not take notes, as he believed it would be distracting.

---

[79] Freud, *Outline*, 26, in the *Penguin Freud Reader*. Independence is the translation for *Selbständigkeit*. The *Standard Edition* translates it as autonomy, which seems to be a reasonable synonym.

[80] Antisemitism, as well as objections to his radical ideas, played a part in his professional appointments to the university. He was *privatdozent* from 1885–1902, which means he was not salaried by the university, but could collect fees from students. After being passed over several times, he was appointed professor extraordinarius (an enhanced *privatdozent*) in 1902.

[81] Kaufmann, *Goethe, Kant, and Hegel*, 15–17. Sigmund Freud, *An Autobiographical Study*, trans. and ed. James Strachey (Norton, 1952), 7. First published in 1935. In *Independent Judgment*, 120–25, I contrast Kantian and individualistic autonomy. I had read Kaufmann on Goethean autonomy years earlier but forgot it. I would like now to give Goethe, as discussed by Kaufmann, credit for my understanding of individualistic autonomy. On the notion of personal autonomy as the goal of psychoanalytic therapy, see Thomas Szasz, *Ethics of Psychoanalysis: The Theory and Method of Autonomous Psychotherapy* (Syracuse University Press, 1965), 16. The term compact majority is from Ibsen's play *The Enemy of the People*.

[82] "Sigmund Freud's Famous Psychoanalytic Couch," *Freud Museum London*, freud .org.uk.

Free association? *Einfall*, according to Bettelheim, a word that means sudden or chance idea or thought, would lead Freud to ask, "What comes to mind?" or "What is connected with that?" *Freie assoziation* were words used by Freud only after the technique had become established and then *Einfall* was described and translated into English as such. The modifier "free" or *freie*, Bettelheim writes, puts too much emphasis on the need for conscious effort, rather than emotional spontaneity coming from the unconscious.[83]

The fundamental rule of psychoanalytic therapy, as Freud refers to it, is that nothing is off the table for discussion and, as he emphasizes, it is not at all a "secular confession," meaning Freud was not there to listen to the patient's sins and provide absolution or condemnation. Whatever connections come to mind, including, especially, painful, nonsensical, or supposedly unimportant thoughts must be expressed while on the couch. This includes slips, reactions to jokes, and the content and memory of dreams.[84]

The Little Hans Case as Illustration of Psychoanalytic Therapy

Freud wrote several case histories to exhibit his method of therapy, as well as his theory of the causes and cures of neurotic symptoms. Let me illustrate this with a summary of his "Little Hans" case about a five-year-old boy with a phobia of horses. The case was published in 1909.[85]

Though Freud only met the boy once, when he was not quite five, the case is a notable example of why Freud thought psychoanalysts do not need to have an MD degree to practice psychoanalysis, contrary to many of his medical critics, including in particular American physicians.[86]

---

[83] Bettelheim, *Freud and Man's Soul*, 94–96.

[84] Freud, *Outline*, 28–29, in the *Penguin Freud Reader*.

[85] Sigmund Freud, "Analysis of a Phobia in a Five-Year Old Boy (Little Hans)," 167–306, in *Case Histories: Dora and Little Hans*, trans. and ed. Alix and James Strachey, present volume ed. Angela Richards (Penguin Books, 1977).

[86] From 1926–32, American psychoanalysts succeeded in getting a law passed in New York state making it illegal for anyone to practice psychoanalysis without the MD degree. Bettelheim, *Freud and Man's Soul*, 33–34. Today, a master's degree is the minimum, but a degree with "doctor" in it is still preferred. In 1926 Freud wrote *The Question of Lay Analysis*, trans. and ed. James Strachey (1950; repr. Norton, 1970), to argue why laypersons—non-doctors—should be allowed to practice his method and theory.

The parents in this case knew Freud. The father of Hans, though not a medical doctor, attended Freud's Wednesday night group (which in 1910 became the Vienna Psychoanalytic Society) and Hans's mother had been Freud's patient. They were both encouraged by Freud to use minimal coercion in their parenting and neither laugh at nor bully Hans about his phobia. The father conducted most of the analysis by asking non-judgmental questions, taking Hans's answers seriously, and sending letters with the answers to Freud. This occurred over the first five months of 1908 before and after Hans turned five.

Hans's main symptom was a phobia of horses, which began as an anxiety of unknown origin while walking in the street with his father. At the time he wanted to go home to "coax" (cuddle) with his mother. On another occasion while walking with his mother, he became afraid that a white horse would bite him. Upon questioning by his father, Hans recalled an event some time earlier when he witnessed a large work horse fall and "make a row" with its feet. This frightened Hans considerably as he thought the horse might be dead. Subsequently, in early 1908, Hans was hesitant to go out into the busy street in front of his home where many horse-drawn carts and coaches would be seen. His anxiety had developed into a full-blown phobia of horses.

The emotions of anxiety and fright that became the phobia are displacements from the repressed cause, according to Freud. Let me now explore the possibilities. There were several.

From the age of three, Hans, like most boys his age, became interested in his "widdler" (penis, or *wiwimacher* in the German) and "widdling." He also naturally became fascinated by the widdlers of others, including those of animals, and curious about whether his baby sister and mother had widdlers. He was especially fascinated by the large widdlers of giraffes and horses. He liked touching his.

At about three-and-a-half, his mother (apparently reverting to older methods of parenting) saw him in this activity and told him that if he keeps touching his widdler she will call the doctor to have it cut off. Because it was implied by his mother that this "activity" was naughty and dirty, Hans acquired what Freud calls a castration complex with the fears becoming repressed. This was also about the time his sister Hanna was born, of whom he soon became jealous.

Other issues were relevant. His parents, unfortunately, told him that his baby sister was delivered by stork, which led to speculation of how that happens. More particularly, when it became obvious that babies somehow come from the mother, Hans wondered how. From the mom's anus, for example?

Freud was critical of the parents for not informing Hans about the differences in sex organs of boys and girls and for not enlightening him about sexual intercourse between his mother and father. This last is important for Freud because he found that children who witness or hear their parents having sex usually conclude that it is an act of violence with the dad harming the mother. Freud concluded that if the parents had been more open about these two issues, the phobia of Hans could have been resolved more quickly. According to the father, Hans never witnessed his parents having sex—although he did sleep in their bed until he was four.

The central issue for Hans revolves around what Freud calls the Oedipal complex, which does not mean that five-year old boys want to have sexual intercourse with their mothers. It is metaphor summing up common issues that arise between children and their parents, especially when there is an attachment to the mother and lesser feeling toward, and in some cases alienation from, the father. The Oedipal myth is about self-discovery, which is what psychoanalytic therapy is about.[87]

Hans did show more attachment to his mother than to his father, but Hans's dad was unusual—for dads then and now—by being so lovingly interested in and patient to help Hans with his phobia, or "nonsense," as Hans referred to it, talking to him about widdlers, babies (his sister), large and small animals (and their widdlers), and often encouraging him to go out to the street where horses could be seen to test his "nonsense." That Hans was aware of his nonsense as an issue indicates his eagerness to seek a remedy.

Sometimes, when the family was away from Vienna on summer vacation, the dad would have to return to the city for business. Hans once said, "When you're away, I'm afraid you're not coming home," which was then followed by a benevolent emphasis by his father that he always comes home.[88]

---

[87] Bettelheim, *Freud and Man's Soul*, 30. Chapter 5 in Bettelheim presents the myth and how Freud uses it in therapy.

[88] Freud, "Little Hans," in *Case Histories*, 206–207.

The attachment to his mother is shown as follows. The "seduction" scene, as Freud refers to it, using the word broadly, not literally, occurred when Hans's mother was giving him a bath and he said to her, "Why don't you put your finger there?" (on his widdler). His mother responded that that would be piggish and improper. On another occasion, which Freud describes as a second seduction, Hans said to his mother that his aunt recently said that he has a "dear little thin-gummy." And once after an anxiety dream, Hans awoke crying, "I thought you were gone and I had no Mummy to coax with." In addition to these issues of attachment to his mother, Hans was not allowed to see her during her confinement (last several weeks) of pregnancy. Much of this came out during the Q&A sessions with his father.[89]

In early April, 1908, Hans and his father paid a brief visit to Freud. Observing the two sitting opposite him and listening to Hans describe how he disliked the black thing around the horses' mouths (the bridle strap above the nose that usually holds a metal bit for control), Freud made a connection between the bridle and the father's black mustache. Freud said that because Hans is so fond of his mother, he must think his father is angry with him. The father responded by asking why Hans would think such a thing, since he, the father, had never scolded or hit him. But yes, Hans intervened, you did hit me once, which happened as a reflex hit after Hans had head butted him in the stomach![90]

After this short visit with Freud, the father reported that Hans's phobia began to retreat, as Hans was more open to talking about his repressed wishes. Many additional concrete details including associations between his feelings toward horses and his feelings toward his parents followed, including more intimate Q&A sessions between Hans and his father, such as: Do you wish your sister had not been born? And on a different occasion, do you wish she would fall into the bath water and die? Both answers were yes, the psychological meaning being that his sister was robbing him of his parents' love. And such an occurrence (her death) would leave more time for him to spend with his mother.[91]

These verbalizations about his sister, father and mother, and his widdler (the castration complex) constitute the ideation Freud talks about that led to a gradual melting of his symptoms of anxiety and fright of horses.

---

[89] Freud, 186–88.

[90] Freud, 203–205.

[91] Freud, 229–35.

Near the end of this period, Hans experienced and talked about a number of fantasies with his father, including his wish to marry his mother and, sometimes, his wish that his father would go away. One day, Hans's father saw him playing with imaginary children and said to him that boys cannot have children to which Hans responded, "I know. I was their Mummy before, *now I'm their Daddy*." Who's their Mummy? the father asked. "Mummy, and you're their *Grandaddy*."[92]

To this, Freud concluded, "The little Oedipus had found a happier solution than that prescribed by destiny"—for Oedipus the King in classical literature. Hans's father wrote to Freud shortly afterwards, "A trace of his disorder still persists, though it is no longer in the shape of fear."

It was the content of the Q&A sessions with the father, as well as the comments made by Freud during the visit with him, that constitute the ideation associated with Hans's emotions, the emotions that troubled him sufficiently to transform them into a fear of horses. And it was this ideation that Freud asserts is necessary to be made explicit and conscious (which in an adult would likely have long since been deeply repressed). The various expressions in Hans's words of his fears and anxieties were the "abreactions" Freud refers to as being essential for successful therapy. The job performed by Hans's father got Hans to bring out or "work off and get rid of" the ideas and emotions of the triggering events, overcoming resistance to talking about them, which then encouraged the displaced emotions in the symptoms to melt away.[93]

FREUD THE SCIENTIST AND HUMAN NATURE

Using Ayn Rand's file folder metaphor, Freud packed the concept of soul—consciousness—with a large number of other universal concepts, both new and revised, to identify healthy and unhealthy motivation and

---

[92] Freud, 256, Freud's italics.

[93] Hans's father was Max Graf (1873–1958), a Viennese musicologist. Hans was Herbert Graf (1903–73), a highly successful opera producer and director, a career that included twenty-four years between 1936 and 1960 at the Metropolitan Opera in New York City. At age nineteen, Herbert visited Freud, who described him as a "strapping youth" who seemed to have survived "puberty without any damage" and, though his parents had divorced and remarried, kept his emotional life intact while going through "one of the severest of ordeals" (the divorce). Freud, "Little Hans," in *Case Histories*, 304–05. See four-part interview of Herbert Graf by Francis Rizzo, "Memoirs of an Invisible Man," *Opera News*, February 5, 12, 19, and 26, 1972, freud2lacan.com.

behavior, and to treat the unhealthy. He did it all without conducting a single positivist-influenced controlled experiment or statistical correlation. Reason was his tool and conceptualization was his method of scientific study. His data came from his patients, his new and revised concepts came from his mental effort to generalize findings to explain the causes of mental health and neurosis, then to apply those new and revised concepts to treat each individual who came to see him.

In the present chapter alone, I have described over twenty concepts, all of them abstractions from abstractions, many of them concepts of consciousness. Among these concepts, not to exclude or ignore the importance of the others, are the topological structure of the soul, its dynamic interactions, and the quantitative (economic) nature of its emotional energy; the definition of repression as a response to anxiety, not a cause of it; and the identification that both psychology and physics are sciences that look for causes of events and methods of influencing those events, even though the subjects of these sciences (molecules and the unconscious) are only inferred, rather than perceived directly by our senses. Which is to say that all science, whether physical or human, is conducted and established through conceptualization.

So what are Freud's contributions to psychology? Nearly everything—from the nature of the conscious and subconscious mind to the cause of neurosis and its treatment in therapy. Talk therapy today is the dominant technique of psychotherapists (unless, unfortunately, they are the psychiatrists who prescribe psychotropic drugs to allegedly treat and cure psychological problems). What else would therapists do but talk with their patients? What nearly all today do not do is mention the words introspection, free will, or the sub- or unconscious, the former two because they fear accusations of being "unscientific" or even religious, the latter, unless they are psychoanalysts, of being accused of Freudianism.[94]

Over the years, ideation has become more explicit as thinking errors in cognitive psychology and in the work of Yochelson and Samenow on the criminal personality. It has become more precise as core evaluations in the psychology of Edith Packer.[95]

---

[94] Cf. William Glasser, *Choice Theory: A New Psychology of Personal Freedom* (Harper, 1998), 4, who refers to psychotropic medicine as "brain drugs" and their alleged side effects as just effects, meaning there is nothing secondary or side about them.

[95] See especially Samuel Yochelson and Stanton Samenow, *The Criminal Personality: A Profile for Change* (1976; repr. Jason Aronson, 1993); Samenow, *Inside the Criminal Mind*; and Packer, *Lectures on Psychology*, chap. 1. Cognitive behavior

As for the supposed weak ego—*the I*—let me quote Freud in *Inhibition, Symptom, and Fear*. In this work, Freud points out the "numerous voices" (other psychoanalysts) who want to make the so-called weakness of *the I* one of the "central pillars" of psychoanalysis. Freud responds: "Shouldn't their sheer awareness of how repression actually works deter psychoanalysts in particular from so enthusiastically embracing such an extreme and partisan position?" The whole point of therapy being to reclaim *the I* and support it. In an earlier paper, Freud explains why *the I* feels alienated by psychoanalysis and may not be the "master in its own house"—namely, because therapy requires the patient to look inward to discover why and how he or she brought about the neurosis. This, says Freud, becomes a "blow" or insult to the patient's narcissism (self-love or self-absorption).[96]

What does impinge on and reduce the power of the conscious mind to control our lives is determinism. Freud's search for non-volitional causes of behavior, I believe, is what led him to focus heavily on the drives, rather than ideation, and specifically on the alleged death or destructiveness drive to explain the harmful and destructive behavior of his patients. Though the notion of a strong unconscious has a long history, going back to Plato, free will is what enables us to learn how to introspect to explain and change our feelings deriving from an un- (or

---

therapy, on the other hand, has been criticized by psychoanalyst Jonathan Shedler as superficial and, in effect, therapy by manual, in "That Was Then, This Is Now," 417–18, 428. "Thinking errors," as an intellectual statement, are the fundamental cause of psychological problems. The therapist at work, however, as discussed by Shedler, does not preach to the patient about what caused the problems and must be done to feel better. Psychological problems have multiple causes, collections of "thoughts, feelings, and behaviors we disavow, repudiate, or defend against." Shedler, 425. The therapist and patient together must carefully work through the past to determine what might be changed in the present. It is a slow process.

[96] Freud, *Inhibition, Symptom, and Fear*, in *Beyond the Pleasure Principle and Other Writings*, 162. The earlier paper is "A Difficulty in the Path of Psychoanalysis," trans. Joan Riviere, in Sigmund Freud, *Psychological Writings and Letters*, ed. Sandor L. Gilman (Continuum Publishing Company, 1995), 269–76. First published in 1917. Leonard Peikoff, apparently accepting most of the conventional criticisms of Freud, unfortunately describes Freud's view of homo sapiens as an "ordure [excrement]-strewn pervert." And Ayn Rand refers to Freud as one of Europe's "hand-me-downs." I am sorry to say that I question whether either read much, or any, of Sigmund Freud. Peikoff in particular derisively dismisses Freud, citing the above common misunderstanding of the supposed weak *I* as evidence. Leonard Peikoff, *The Ominous Parallels: The End of Freedom in America* (Stein and Day, 1982), 213–14. Ayn Rand, "Don't Let It Go," in *Philosophy: Who Needs It* (Bobbs-Merrill, 1982), 256. First published in *The Ayn Rand Letter* 1, nos. 4–5 (1971).

sub-) conscious "chaos" or "cauldron full of seething excitations," as Freud describes the unconscious. The chaos is something we had some part in putting there, which means we can, through introspection, clean up the mess.[97] With improved parenting and teaching, one would hope, psychology of the future will help prevent the mess from occurring in the first place.[98]

Freud's theory of human nature is summarized in his widely read and influential work, *Civilization and Its Discontents*, whose title Bettelheim insists would have been better translated as "The Inherent Uneasiness in Culture."[99] In the absence of free will, the drives, especially the death and destructiveness drive, give us tendencies to aggression. The aim of culture is to curb those tendencies, thus producing the "inherent uneasiness" within. A few of us, though, may sublimate the drives. Interestingly, in this work, Freud expresses his opposition to communism and defends private property. However, he is not too approving of American culture.[100]

A Theory of Human Nature

The significance of free will for a theory of human nature is that human beings are born neither good nor evil. "We choose our own misery," so says psychiatrist William Glasser (paraphrased). "We can choose our own happiness."[101] Which means we can go either way, but from the standpoint of physical and mental health, I would say that we must modify this "neutral" neither-good-nor-evil theory. Just as the body has a built-in tendency to heal itself, though sometimes fails, so also does the mind. I call this tendency the "will to do the right thing."

---

[97] Freud, *New Introductory Lectures*, 91.

[98] Cf. Kirkpatrick, *Montessori, Dewey, and Capitalism*, esp. chap. 4, and *Independent Judgment*, esp. 179–80.

[99] *Das Unbehagen in der Kultur* in the German. *Kultur* emphasizes moral values and intellectual and esthetic achievements. Bettelheim, *Freud and Man's Soul*, chap. 14. *Zivilisation*, by contrast, emphasizes the material accomplishments of society.

[100] Sigmund Freud, *Civilization and Its Discontents*, trans. Ulrich Baer (Warbler Press, 2022), 59–62, but chap. 5 generally. First published in 1930. Freud's disapproval of American culture, I assume, is because of our emphasis on material values. He also did not like Woodrow Wilson.

[101] Glasser, *Choice Theory*, chap. 1. Glasser was trained in psychoanalysis but shifted his practice away from what he called "insight therapy" to changing his patient's present behavior, hence his emphasis on choice and choosing better behaviors. See Kirkpatrick, *Independent Judgment*, 19–21.

What we will to do does not always mean doing the morally or psychologically right thing; we can after all still make mistakes or even do evil. However, we see this desire for rightness or goodness in children from their earliest years, eagerly seeking to live their lives in a healthy and happy way, often continuing through adolescence and adulthood after being confronted with major, unpleasant environmental obstacles.[102] These children and adults still differ according to their respective parental and educational upbringings, that is, their general knowledge, and the particular values and consequent experiences they have chosen to pursue, their personal knowledge.

What doing the right thing means, more specifically, is doing what we *perceive* or *believe* to be right. This is a psychological and biological need to *feel* right before taking action. Even criminal personalities have this need to *feel* right, which explains the prevalence of rationalization as their defensive habit to justify crime. The will to do the right thing can be described as a "perceived rightness" that varies from person to person, analogous to the psychological principle of perceived risk, which means the risk we are willing to take varies from person to person.

These comments give us a nuanced theory of human nature that means our character and personality is self-made, for good or ill, but given the right circumstances and choices we will tend to lean toward rightness or goodness. The theory recognizes that human beings, because they possess the capacity to reason, are capable of making rational decisions, especially, as one important example, those required to raise themselves above their original stations in life—through education, occupation, and income—in a free society, namely a political and economic system called capitalism.

In civilized cultures, to extend this point further, I would say that *most* human beings choose to be essentially honest and to exhibit a benevolent intention, differing only in degree of understanding of what precisely it means to be honest. By working hard and actualizing their potential to reason, that is, to *initiate reason*, rather than initiating coercion, humans seek to cooperate with others to improve their lives. Only a minority, criminals, choose to "enhance" their lives at the expense of others, and may do terribly evil things.

---

[102] We can also observe this eagerness to live in the higher mammals, newborn puppies and kittens, for example—to seek food, to enjoy playing with their littermates, to learn to bark or meow, and to hunt. Sometimes, they may walk over the runt of the litter, but not always.

When I say "most human beings are essentially honest," I am talking primarily about laypersons, who in civilized cultures are the majority. Whether they have been influenced about honesty by Immanuel Kant or Jeremy Bentham or Ayn Rand or Jesus, I submit that their intentions are benevolent. Some political leaders of a culture and their intellectual influencers, as we can read about in history, may or may not be benevolent.

The significance of parental and educational upbringing puts a premium on a sound education, not just of children, but also, or more importantly, of the adults who are their children's teachers.[103] Human behavior is learned, but from the time we are born we have a built-in urge to live and thrive—though not quite the same as Freud's libido drive and definitely not the destructiveness one.[104]

Freud, Individualism, and Capitalism

Freud did not have an explicit epistemology, especially a theory of universals, and like many psychologists then and today he did not explicitly discuss the connection between the individualism of psychology and capitalism. Individualism is a moral and political concept that presupposes the above nuanced theory of human nature. It rests on the notion that the primary units of reality are individual concrete entities, including, especially, individual human beings who each possess a volitional, reasoning mind that must be put to use to identify and accumulate knowledge of reality. Based on that accumulated knowledge, each of us as individuals, who possess the inalienable rights to life, liberty, and property, choose and act to pursue our own values, working with others through mutual, voluntary agreement and trade. Collectivism is the opposite.

---

[103] "The problem of how better to relate to the child is one of education—of the adult." *Montessori, Dewey, and Capitalism*, 133.

[104] For an expanded explanation of this theory, see my blog post, "On the Nature of Human Nature," *Jerry Kirkpatrick's Blog*, July 22, 2025, jerrykirkpatrick.blogspot.com. Cf. Rutger Bregman, *Humankind: A Hopeful History* (Little, Brown, 2019), for his positive view of human nature, illustrated by an abundance of examples, including the story of the six Tongan boys stranded for fifteen months on the uninhabited island of Ata. *Lord of the Flies*, the title of William Goldman's fictional tale of stranded boys, observes Bregman, it was not. Bregman, unfortunately, does not share my view of capitalism, apparently preferring a socialistic form of "participatory democracy."

Humankind, or society, is not an entity; it is a collection of individuals, as are groups and masses. Ethics focuses on the moral values and virtues of the individual. Political philosophy and economics both must then look at the behavior of individuals in social relations to define, respectively, the proper functions of government and the requirements for producing wealth.

Freud said he was a "liberal of the old school," which means he advocated classical liberalism, the doctrine that emphasizes the individual rights of freedom, equality before the law, and private property, requiring a limited government.[105] Freud also insisted on the primacy of the individual in psychology, namely that individual psychology must precede any discussion of social or mass psychology.[106]

The reason we study psychology before economics and other human sciences is that we need to understand individual motivation and behavior—both normal and abnormal—before we can recommend choices and actions in an interacting social setting.

We turn now in chapter 5 to economist Ludwig von Mises who explicitly discusses epistemological issues in order to provide a foundation for his work in economics, and who implicitly assumed a sound theory of psychology, especially the work of Freud. Mises is the foremost economic defender of the free society.

---

[105] "In spite of all my dissatisfaction with the present economic systems I have no hope that the road pursued by the Soviets will lead to improvement. Indeed, any such hope that I may have cherished has disappeared in this decade of Soviet rule. I remain a liberal of the old school." Freud to Arnold Zweig, November 26, 1930, 21.

[106] Freud, *Mass Psychology and Analysis of the I*, chap. 1, in *Mass Psychology and Other Writings*, trans. J. A. Underwood (Penguin, 2004). First published in 1921.

5

Ludwig von Mises as Scientist

> The human mind is not a tabula rasa on which the external events write their own history. It is equipped with a set of tools for grasping reality. Man acquired these tools, i.e., the logical structure of his mind, in the course of his evolution from an amoeba to his present state. But these tools are logically prior to any experience.
>
> —Ludwig von Mises[1]

> Value is not intrinsic, it is not in things. It is within us; it is the way in which man reacts to the conditions of his environment.
>
> —Ludwig von Mises[2]

FOR OVER TWO CENTURIES, the scientific study of human beings has been given various names, often followed by the word "science": moral, human, humanistic, social, and cultural. Other suggested names have been the science of human action, or just the humanities.[3]

---

[1] Ludwig von Mises, *Human Action: A Treatise on Economics*, scholar's ed. (Ludwig von Mises Institute, 1998), 35. German edition published as *Nationalökonomie* in 1940. First English edition, rewritten in English and edited by the author, published in 1949. The scholar's edition is essentially this first edition. Subsequent references to the Ludwig von Mises Institute as publisher will be listed simply as Mises Institute.

[2] Mises, *Human Action*, 96.

[3] The German word *Geisteswissenschaften* is usually translated as "humanities," though sometimes, because *Geist* means "mind," it is translated as "mental science," which I prefer because "mental science" is broader than humanities and is a close synonym of "human science." None of these terms should be confused

Unlike Sigmund Freud, Austrian economist Ludwig von Mises explicitly discusses epistemology to provide a foundation for his theory of economics. In his lengthy but unappreciated career, Mises in the twentieth century was unsurpassed in his economic defense of the totally free society, that is, laissez-faire capitalism. Like Freud, Mises out of necessity left his homeland to escape the Nazis, immigrating to the United States in 1940.[4] From that time on, he continued to teach and write in English, attracting a large following of students in the United States.[5]

As I did with Freud in chapter 4, I will review the work of Mises, his basic and applied science, including his approach to science—conceptualization—in the context of Ayn Rand's theory of concepts. In the first section of this chapter, I will define and delimit the epistemological meaning of the science of homo sapiens and differentiate it from two terms in use today: social science and the preferred term of Mises, praxeology, the science of human action. In the second and subsequent sections, I will discuss the explicit epistemology of Mises, especially his concessions to Kant and what I will argue are his epistemological errors. I will follow up the discussion of epistemology with a presentation of Rand's value theory in relation to economics in general and to Mises in particular, as well as to ethics. I will conclude the chapter by

---

with academic departments in universities or the courses they teach, as these departments and courses usually represent more specific subjects of study determined by current fashion.

[4] Freud escaped the Nazis by moving to London in 1938. A daylong interrogation of his daughter, Anna, by the Gestapo convinced him it was time to leave. Gay, *Freud*, 625–26.

[5] By "unappreciated career," I mean the following. Only briefly did Mises enjoy a salaried position paid for by a university, from 1934–40, at the Graduate Institute of International Studies in Geneva, Switzerland. From 1913–18, Mises was *privatdozent*, or private lecturer, at the University of Vienna, which meant he, like Freud, could collect fees from his students. From 1918–34, he was professor extraordinarius, an advanced, tenured *privatdozent*, but not a full, salaried professor with a chair. From 1948–69, he was permanent "visiting professor" at New York University with his salary paid initially by a nonprofit organization and, later, by a New York businessman. Meanwhile, socialist émigrés from Nazi Europe were given prestigious chairs at American universities. Mises earned most of his income in Vienna as economic advisor to the Austrian Chamber of Commerce, 1909–34, part-time until 1938. Similar to Freud, antisemitism and opposition to his radical ideas thwarted Mises from receiving due recognition in his home country. See Mises biographies by Murray Rothbard, *The Essential Ludwig von Mises* (1973, 1988; repr. Mises Institute, 2009) and Jörg Guido Hülsmann, *Mises: The Last Knight of Liberalism* (Mises Institute, 2007).

highlighting Mises's basic and applied science of economics. Whatever errors I attribute to Mises should not be understood as diminishing his accomplishments and heroism in the twentieth century.

HUMAN SCIENCE VERSUS SCIENCE OF HUMAN ACTION

Which is it? Human science or science of human action?

To provide a broader context, here is a quotation from my book *In Defense of Advertising*, elaborating a point I made in the first chapter of the present work:

> It has been said that there are three fundamental sciences: physics, biology, and psychology, or physical science, life science, and human science. . . . Another way to describe the division is: (1) the sciences of inanimate matter; (2) the sciences of the life forms that possess a vegetative function of life or, in addition, the conscious function of sensation, or, still further, the conscious function of perception; and (3) the sciences of the life form that possesses, in addition to the vegetative functions and the conscious functions of sensation and perception, a volitional, conceptual consciousness.[6]

"Note that the sciences," I go on, "are cumulative [and overlap] in the sense that principles of physics and chemistry and of the vegetative actions of life are used in the life science of human medicine in such specialties as biomedical engineering, molecular medicine, and cardiology."[7] Physical sciences study inanimate matter; the life (or biological) sciences study living organisms, and the human sciences study human beings.

The aim of the human sciences is to study and develop a theory of human nature as the rational animal that possesses a volitional, conceptual consciousness, which means one that exhibits the *capacity* to reason, not one that reasons automatically. The focus of human science is on the motivation and behavior primarily of the individual and only secondarily of the individual in a group. This makes psychology the basic, foundational field of study of the human sciences.

---

[6] Kirkpatrick, *In Defense of Advertising*, 162n44. See above, pp. 24–29.

[7] Cf. Harry Binswanger, *The Biological Basis of Teleological Concepts* (Ayn Rand Institute Press, 1990), 4–16.

The term "social science" has been around at least since the early nineteenth century but became dominant in the twentieth. It refers to the study of human beings in society and is a close synonym of "sociology," a word coined by the collectivist philosopher Auguste Comte and defined as the science of society.[8]

Comte also coined the word "positivism" as a post-Kantian attempt to identify "true reality," which Kant of course said we cannot know, by restricting awareness to directly perceived concretes, all else being empirically meaningless with any abstract concepts we come up with said to be "arbitrary constructs." In subsequent elaborations positivism dismissed metaphysics, universals, objective facts, and objective values, and truth, all as unknowable. All we can do is mimic the physical sciences in method and produce statistical probabilities.

The emphasis on "social," not individual, human beings and statistics as method, because it looks at groups to come up with averages and percentages, made this so-called science of human beings collectivist at root. "Society," in other words, became the unit of analysis or entity of study, not the individual.

Over the years, use of the word "social," or the prefix "socio-," proliferated to modify many disciplines and endeavors: social psychology, social engineering, social justice, sociobiology, socioeconomics, sociology of (physical) science, and so on. Because of this collectivization of human science, I refused in *In Defense of Advertising* to endorse the word "social" when writing about the "social" criticisms of advertising. "It is a misnomer," I wrote:

> to refer to these charges [against advertising] as "social" criticisms, for the term "social" implies that morality is essentially a social concept. It is not. Morality defines a code of values to guide each individual in his choices and actions. Hence, the quotation marks around "social."[9]

Individualism and the individual are primary. As Freud biographer, Peter Gay, puts it, Freud "was fully persuaded that individual and social

---

[8] "Social science" is not the same as the branch of philosophy variously called social philosophy, political philosophy, or, simply, politics. Political philosophy rests on and applies the concepts and principles of its underlying branches, metaphysics, epistemology, and ethics, to study the nature and proper function of government.

[9] Kirkpatrick, *In Defense of Advertising*, 34n12. I argued that the so-called social criticisms, namely that advertising is coercive and offensive, are at root philosophic.

psychology are impossible to separate."[10] Not just impossible to separate, I would add, the individual is the starting point of analysis, proceeding later to the individual's relation to and interaction with others in a group setting. Society is an assortment or assemblage of individuals who are the primary entities.

In the late nineteenth century, the *methodenstreit*, or dispute over methods, erupted between Carl Menger of the Austrian School of economics and Gustave Schmoller of the German Historical School. Menger advocated a distinctively Aristotelian theory of universal principles and their application to *individual* human action. Schmoller rejected such an "essentialist" epistemology and "methodological individualism," and advocated the positivist notions of collecting large quantities of historical data, especially statistics.[11]

The dispute pointed out to the Austrian economists that economics was a narrower subject of a broader theory of human science. Ludwig von Mises, at first, accepted "sociology" as the parent discipline, but rejected its collectivist premises and followers. He settled on praxeology—a word not coined by him—to define a science of human action, or "action as such," as he puts it. Many of his followers today continue to base the foundation of economics on praxeology.[12]

When I first read Mises, I found his discussions of praxeology thorough and intriguing, but not consistent with my understanding of Ayn Rand's epistemology. My conclusion today is that Mises was caught up in German philosophy such that, though he went a long way toward rejecting many of its faulty premises, he failed to remove all the primacy-of-consciousness notions, most notably his acceptance and use of Kant's words "a priori categories." Mises's use of these terms, however, is more Aristotelian than Kantian, which I will get into in a later section.

The problem with a science of human action is that it lacks an explicit theory of human nature. As a concept it tries to account for all human behavior in what amounts to a substitute either for moral or practical

---

[10] Gay, *Freud*, 338.

[11] See Samuel Bostaph, "The Methodological Debate Between Carl Menger and the German Historicists," *Atlantic Economic Journal* 6, no. 3 (1978), 3–16, https://doi.org/10.1007/BF02313305.

[12] The term "praxeology" was coined as early as 1882 from the Greek word *praxis*, meaning action or practice. See Jörg Guido Hülsmann, Introduction to the Third Edition of *Epistemological Problems of Economics* by Ludwig von Mises, 3rd ed., trans. George Reisman (Mises Institute, 2003), xxii. First edition published in German in 1933.

philosophy, without addressing or accepting the metaphysics—the philosophical psychology—of a volitional consciousness. At the same time, it studies all human choices, while denying that it has anything to do with ethics or psychology.[13] Nonetheless, Mises abundantly uses such terms as choice and volition (caused by ideas), value judgments, and purposive behavior (of determining the means by which to achieve ends), but does not explicitly endorse free will, leaving that issue to other (physical and biological) sciences as an unanswered problem of determining the relation between our brains and consciousness. All we can know, says Mises, is that an acting person's choices derive from "his individuality—the product of all that he has inherited at birth from his ancestors and of all that he himself has experienced up to the critical moment" of making a final choice, which is genes-and-environment determinism.[14]

Praxeology as a concept ignores ethics because the science of "action as such" is said to be "formal" and "value-free,"—neutral, in other words—yet the production of wealth as the aim of economics is hugely beneficial, that is, a value, to human life.[15] Further, praxeology does not recognize or accept human psychology as the basic, foundational human science. In its study of motivation and behavior, however, psychology includes the study of "action as such." Mises, using a positivist premise, relegates the basic science of psychological

[13] Cf. Robert Mayhew, ed., *Ayn Rand's Marginalia: Her Critical Comments on the Writings of over 20 Authors* (Second Renaissance Press, 1995), 117, 133–34, and Roderick Long, "Praxeology: Who Needs It," *The Journal of Ayn Rand Studies* 6, no. 2 (2005), 299–316.

[14] Ludwig von Mises, *Ultimate Foundation of Economic Science* (D. Van Nostrand, 1962), 58. See also Mises, *Human Action*, 46–47. Freud essentially believed the same thing, namely that science would one day conclusively demonstrate the connection between our brains and consciousness, proving that free will is an illusion. No doubt, science one day will discover some kind of connection between our brains and consciousness, but that will not deny the validity of free will, as the scientists will be using their free will in the process of discovery. The science will more likely demonstrate how a volitional consciousness arose in the process of evolution. See Damasio, *Feeling and Knowing*, 31, 109–110, for his discussion that feeling, as a "hybrid of mind and body," arose in evolution before consciousness, about 500 million years ago with the arrival of nervous cells and primitive nervous systems. See also Norman Doidge, *The Brain that Changes Itself* (Penguin, 2007), esp. chap. 5, on the neuroplasticity of the brain and for stories of human consciousness—through will power—to change its brain and improve and cure paralyzed legs and arms.

[15] More later on the production of wealth as the meaning of economics. Mises does not use that definition.

theory to experimental psychology, a naturalistic psychology, as he calls it, and the applied science of psychotherapy, which he prefers to call thymology, to history.[16]

Thymology, like praxeology, was not coined by Mises. The term derives from the Greek word *thumos*, meaning soul or spirit, or feeling or thought. "Thymology" is history in the sense that the psychotherapist must understand the patient's concrete past and present thinking errors that are causing the unhappiness, as similarly the medical doctor must identify the specific ailment of a specific patient to recommend treatment. Both psychotherapy and medicine rely on a vast store of previously learned universal concepts and principles that apply to these cases.

Application, as I have written earlier, is the epistemological, deductive process of conceptualization and essential part of any applied science.[17] Generalization is the prior, inductive component. "Thymology," I contend, is not a necessary or helpful concept, because applied psychology is the applied human science that psychotherapists practice, and basic or theoretical psychology is the inductive, identification-of-concepts science that is broad enough to include the thinking of experimental practitioners.[18] Mises, as do I, rejects the pejorative term "literary psychology," sometimes unjustly applied to Freud.[19]

My objections to the concepts of praxeology and thymology reveal the crucial error of a science of human action, namely that it does not have a valid theory of universals to answer the irrationalism of the past two hundred years. Mises correctly recognizes that the epistemological

---

[16] Mises, *Theory and History*, chap. 12 and Mises, *Ultimate Foundation*, chap. 2, sec. 8. In a sense, the concept of praxeology is at once too broad and too narrow, attempting to study all action "as such" (too broad) without bringing in either ethics or psychology (too narrow). True, economists do not usually need to know the specific motivation of consumers or workers, but the applied business sciences of marketing and management do, both of which rest on psychology and economics. See this early paper by me on "Theory and History in Marketing," *Managerial and Decision Economics* 4, no. 1 (1983), 44–49, https://doi.org/10.1002/mde.4090040106. Originally a conference presentation published in Ronald F. Bush and Shelby D. Hunt, eds., *Marketing Theory: Philosophy of Science Perspectives* (American Marketing Association, 1982), 47–51.

[17] See above, p. 29.

[18] See "Rand's razor," *Objectivist Epistemology*, 72, 250–51, namely that concepts must not be multiplied beyond necessity.

[19] Mises understood and praised highly the work of Sigmund Freud and dismissed out of hand all attempts to connect Marxism and psychoanalysis. Mises, *Theory and History*, 27, 152–54, 268, 281.

problems of economics lie in the forms of irrationalism he was up against. These include polylogism (the theory of many logics, which is an attack on Aristotelian logic), historicism (the theory that history is the only source of knowledge in the human sciences, meaning it can only produce time-bound statistical generalities and probabilities), and positivism, all three of which in their various incarnations plague us to this day.

Kant's transcendental idealism told us reason is impotent to know reality. Ayn Rand's theory of concepts, however, through her theory of abstraction as measurement omission and, perhaps more significantly, her explanation of abstraction from abstractions, has given us a theory of epistemology grounded in perceptual reality. Conceptualization, properly understood and practiced, is what enables us to study and understand homo sapiens as the rational animal with a volitional consciousness, which enables us to study and understand human motivation and behavior as an individual and as a member of society.

Psychology, economics, political philosophy, and all the applied human sciences can apply Rand's epistemology to their fields without obsessing continually over exact measurement or probabilities. Human science, as understood using Rand's theory of concepts, not praxeology, is the general theory that provides the foundation for Mises's science of economics.

MISES AND KANT

In his work on the *Categories*, Aristotle identified ten fundamental concepts of reality, for example, substance (entity), quantity, quality, action, and relation.[20] Immanuel Kant describes these categories as fundamental, innate concepts that prevent us from knowing true, noumenal reality. We can only know appearances, says Kant, which he calls the phenomenal world. Hence the impossibility of perceiving "things-in-themselves," which means concept formation is limited and cannot know concretes.

This is what makes reason impotent to know reality and in essence is Kant's disastrous influence on later philosophy. It has confounded nearly all who have followed him, including economists such as Ludwig von Mises.[21]

---

[20] Aristotle, *Categories*, 1b25–15b31, in Mckeon, *Basic Works*, 3–37.

[21] Cf. Heinrich Rickert, *The Limits of Concept Formation in Natural Science: A Logical Introduction to the Historical Sciences*, ed. and trans. Guy Oakes, abridged ed. (Cambridge University Press, 1986), original edition published in German in

To return to the infant's epistemology discussed in chapter 2 of the present work, knowledge of concretes begins at birth. The infant cries when hungry and cries when having to eliminate, meaning the infant has become aware of its mother's breast and its own bodily processes. We know this because infants, after a short time out of the womb, will give us a cute smile when sated and vacated. This is one of the baby's earliest awarenesses of concretes, through sensation. Later, through perception, it becomes aware of many more concretes.[22]

Awareness of the concretes of reality is incorporated into Ayn Rand's theory as the starting point of her theory of concepts. A child of one or two, after learning some elementary words describing concretes, such as "mama" and "Fido," begins to form and apply universal concepts, such as "ball," "dog," and "table," by observing the differences and similarities of the three types of existents. The concepts are formed by omitting their measurable differences, that is, by observing that Fido and Rover are different from each other, but more similar than the considerably different Sylvester the cat, and, especially, the round, spongy things that bounce and roll.

The parents, of course, are the ones teaching the child that Fido is a dog and the round, spongy thing is a ball, but the child's mind is doing the mental work of recognizing these differences and similarities, because, as Aristotle said, the mind is "so constituted as to be capable of this process"—or processes—of integration and retention by attaching a word to the mental work. The concepts so formed, using Rand's file folder metaphor, include all observed concretes past and present, enabling us to identify future concrete existents of a similar type. The (mental) omission of measurements gives us the essential distinguishing characteristic and the nature of the concept as universal. The word labels the concept, and its definition summarizes and condenses its meaning, which refers to its similar referents "out there" in reality.[23]

1895, and Rickert, *Science and History: A Critique of Positivist Epistemology*, trans. George Reisman, ed. Arthur Goddard (D. Van Nostrand, 1962), original edition published in German in 1899. Rickert seems to be wrestling with his Kantian premises, the problem of how we can know the concretes of history and what the role of values are in that process. See below, the section "On Subjective Value," pp. 132–138, for my discussion of Ayn Rand's general theory of value.

[22] See above, pp. 34–36.

[23] Perceived concretes, as Rand points out, are usually defined ostensively and their definitions can be challenging to put into words. Parents define perceived concretes for their children by pointing. Rand, *Objectivist Epistemology*, 41.

This is the inductive process of concept formation. Application is the deductive process of recognizing an object initially unfamiliar to us and identifying it as an instance, a concrete, of the concept.

Now, what can we say about some of Mises's confusions caused by German, especially Kantian, philosophy? Mises, as quoted in the first epigraph of this chapter, refers to a priori categories and the logical structure of the human mind and denies that our minds are tabula rasa at birth.[24] If by "logical structure of the human mind" Mises means that consciousness has a rational, active nature, or identity, then he is correct, but Mises goes on to say that such categories as causality and human action are innate, prior to experience, like, allegedly, mathematics and logic. Therefore, he claims, deduction is the fundamental method of human science.

The distinction, however, between a priori and a posteriori knowledge is false, deriving from Kant's analytic-synthetic dichotomy that insists some knowledge must be innate. All cognition, however, is cognition of reality.[25] If our concepts are true—correctly identifying reality—then the most abstract concepts, such as the law of non-contradiction, causality, human action, the law of supply and demand, and integral calculus, are all derived from perceived concretes in reality and are therefore empirical.[26] The chain of abstractions required to arrive at these concepts may be extensive, but, if true, the concepts will be anchored in the perceptual world, in the same way that space, time, and causality are so anchored, as demonstrated in chapter 3 of this work.[27]

---

[24] Mises, *Human Action*, 35.

[25] See Peikoff, "Analytic-Synthetic Dichotomy," in Rand, *Objectivist Epistemology*, 93–94, 112–18. The analytic-synthetic dichotomy derives from Kant's appearance-versus-reality premise and gives us these other nuanced, but false, distinctions: a priori versus a posteriori, logical versus empirical, and necessary versus contingent. The only valid distinction here is Rand's, "The Metaphysical Versus the Man-Made," in *Philosophy: Who Needs It*, 28–41. First published in *The Ayn Rand Letter* 2, nos. 12–13 (1973). Rand's distinction refers to what we human beings can change (the man-made) and what we cannot (the metaphysical).

[26] Examples of perceived concretes, in order as listed above: non-contradiction: the glass on my desk is a glass, not a non-glass, that is, not a leprechaun or any other entity, nor is it an empty space; causality: the snake with a diamond shape on its back bit me when I stepped on it, which means I must seek medical help; human action: humans, from the time they evolved from apes, have sought to acquire values to sustain and improve their lives; supply and demand: sellers and buyers for millennia have interacted with each other to transact business; and calculus: I need to estimate an area of land bounded by three straight fences and one curved.

[27] See above, pp. 61–64.

Mises does not seem to have a definite notion of abstraction from abstractions, only directly perceivable concretes, an influence of positivism, and the innate categories from which all of economics is supposedly deduced. Economics is a science of application, as I identify it, a "how to" discipline that aims at guiding human beings to peaceably produce wealth. And economists do spend a great deal of time providing policy analysis and recommendations, which requires the application of the basic concepts and laws of economics to specific problems, but the basic concepts and laws have been empirically and inductively discovered from experience. The concept of human action is not axiomatic or self-evident, as Mises asserts. Action is a fundamental attribute of all living organisms, human action of human beings. All life, as Rand defines it, is self-generated and self-sustaining action, and every living organism exhibits goal-directed behavior, to continue its life.[28]

Mises writes, "The human mind is not a tabula rasa on which the external events write their own history." This is true. Reality does not write on our minds. We actively identify it. "[The mind]," Mises continues, "is equipped with a set of tools for grasping reality."[29] This also is true. We have tools, such as perception, conceptualization, and evaluation—mental processes that we have the innate capacity or potential to perform. Those processes constitute the nature or identity of consciousness, but they are not content. Tabula rasa means there is no content in our minds at birth. The infant has no content in its mind at birth, though it does have biological drives, or urges to act, to lead it to seek a breast or to defecate.

Mises further writes: "Man acquired these tools, i.e., the logical structure of his mind, in the course of his evolution from an amoeba to his present state. But these tools are logically prior to any experience."[30]

---

[28] Mises, *Human Action*, 38–39. Ayn Rand, "The Objectivist Ethics," in *The Virtue of Selfishness: A New Concept of Egoism* (NAL Books, 1964), 5–6. Originally, a lecture delivered at the University of Wisconsin, Madison, February 9, 1961. All living organisms—plants, animals, humans—have a kind of "intelligence" or competence, using Damasio's words, to respond to what they have "sensed" or detected in their environments, to continue their lives. Damasio, *Feeling and Knowing*, 13–14. Spiders and ants become aware of threats to their lives if we get too close to them with a big hand or foot. They quickly scurry away if they can. See Binswanger, *Teleological Concepts*, 13–16, on goal-directed versus purposeful behavior.

[29] Mises, *Human Action*, 5.

[30] Mises, 35.

The mental tools are built-in from birth, yes, and we use them to identify, not passively receive or reflect, reality and to guide our choices and actions. Mises is arguing against the mirror-of-nature theory of the mind. The only option he sees, unfortunately, is Kant's innate categories.

The expression "logical structure of [man's] mind," as used by Mises, is confusing. Kant thinks logic is prior to experience and thought. In fact, experience and thought are prior to the use of logic. Abstraction from abstractions, whether we are talking about integral calculus or the law of supply and demand, are derived from perceptual concretes and can be traced to their origin in those concretes.

We must now ask: Does Mises, as a user of some Kantian terms, believe we can know reality? Yes, and this, along with his methodological individualism and emphasis on the nature or essence of things (money, capitalism), makes him Aristotelian, or as Hülsmann says, "a representative of Aristotelian realism."[31] In his last book on epistemological issues, Mises writes, "From the praxeological point of view it is not possible to question the real existence of matter, of physical objects and of the external world."[32]

In *Human Action*, though, he is equivocal: "It is idle to ask whether things-in-themselves are different from what they appear to us, and whether there are worlds which we cannot divine and ideas which we cannot comprehend. These are problems beyond the scope of human cognition."[33]

The quotation apparently is Mises's way of dismissing as unimportant Kant's dilemma of appearance versus reality and of asserting his own emphasis on analyzing the world in which we live—the Aristotelian world. A few pages later, Mises writes, "Praxeology conveys exact and precise knowledge of real things," and in *Ultimate Foundation*, he says, "We see reality, not as it 'is' and may appear to a perfect being, but only as the quality of our mind and of our senses enables us to see it," which I take to mean reality with no Kantian distortions.[34]

---

[31] Hülsmann, Introduction to *Epistemological Problems*, liii. See also David Gordon, *The Philosophical Origins of Austrian Economics* (Mises Institute, 1993) and Barry Smith, *Austrian Philosophy: The Legacy of Franz Brentano* (Open Court, 1994), chap. 10.

[32] Mises, *Ultimate Foundation*, 6. See also Mises, *Human Action*, 39.

[33] Mises, *Human Action*, 36.

[34] Mises, *Human Action*, 39, and Mises, *Ultimate Foundation*, 18.

Mises here correctly identifies the obsession of philosophers who insist on eternal knowledge and even the omniscience of a "perfect being" or "god's-eye" view of the world. There is no "perfect being" or "god's-eye" view of the world. The god's-eye view is precisely what advocates of the doctrine of pure and perfect competition hold up as a standard by which to judge business activity. We do not live in a phenomenal world, unable to know the noumenal. We live in reality, and that appearance of a bent stick in water is a correct perception of what looks like, or rather appears to be, a bent stick, though with the use of our sense of touch and our knowledge of the causal effects of light in water, we know that the "thing-in-itself" stick is not bent.

To put a point on it, Mises writes, "We study science for the sake of real life . . . not as a form of mental gymnastics"—that is, not rationalistically pushing words around, trudging through the weeds and quagmires of academic minutia.[35]

In *Ultimate Foundation*, Mises does offer this hypothesis: the a priori categories, he says, are *not innate ideas*, but are "the necessary mental tool to arrange sense data in a systematic way, to transform them into facts of experience, then these facts into bricks to build theories, and finally the theories into technics to attain ends aimed at." In this way, they provide "information about the reality of the universe." A remarkable, even Aristotelian, statement that sounds like, though not as worked out as Rand's, a tentative theory of concept formation and abstraction from abstractions, albeit lacking a detailed description of how these processes work.[36]

In opposition to Kant (and other intellectuals), Ayn Rand insists that it is not a claim to omniscience to say there is nothing in the universe unknow*able* to the human mind. "Uknow*able*" and "unknown" are not the same concepts. What we do not now know is vast. Reason and the human consciousness, however, given the right conditions, are *capable*

---

[35] Mises, *Epistemological Problems*, 15.

[36] Mises, *Ultimate Foundation*, 16. This kind of statement, to me, is what makes Mises so stimulating to read. When I first read Mises, I often substituted in my mind the words "universal concept" or "universal principle" for *a priori categories*. I later learned to do the same for the *Standard Edition* translations of Freud, substituting "load" or "investment" for *cathexis*. Mises and Rand did know each other. Is it possible that he was influenced by her?

of knowing anything in the universe.[37] The right conditions may require that we build rocket ships to send us to Mars and beyond to identify what we would like to know. Nothing is hidden from us or held secret, known only to the well-educated, such as Plato's philosopher-kings or an omniscient and omnipotent deity.[38]

MISES AND EPISTEMOLOGY

The human sciences aim to explain the causes of human motivation and behavior, healthy and effective, as well as unhealthy and harmful, and to guide humans in their choices and actions to achieve chosen goals. This includes all applied human sciences whose aims are to get things done and done well. Economics is not just a highly deductive science—deriving many of its precepts from the law of supply and demand—but also highly applied in that its aim is to define the principles of cooperation under a division of labor that will secure peace and prosperity in a social setting.

The Specific Understanding and Intuition

To continue with a few additional epistemological issues in Mises's writing, the two fundamental methods of cognition, according to Mises and the Kantian philosophers, are conception and understanding. Concepts, we have seen, according to Kant are limited to the phenomenal world, which means we cannot know true reality or concretes. This poses a problem in particular for historians, since their work is to explain concrete human and natural events of the past. The solution in German philosophy is the specific understanding (*spezifische Verstehen*). Mises writes:

> It is the method which all historians and all other people always apply in commenting upon human events of the past and in forecasting future events. . . . The scope of understanding is the mental grasp of phenomena

[37] Though what we know, to use Freud's word, may be *unerkennbar*, that is, unrecognizable or invisible to the senses. Gravity, molecules, and the subconscious are still known.

[38] On the "unknowable," see Nathaniel Branden, ""Intellectual Ammunition Department: Is there any validity to the claim that certain things are unknowable," *The Objectivist Newsletter* 2, no. 1 (1963), 3.

> which cannot be totally elucidated by logic, mathematics, praxeology, and the natural sciences.[39]

The philosopher Henri Bergson, according to Mises, called this cognitive power "intuition," the unfortunate word used to translate *nous* in Aristotle.[40] Paraphrasing in English Mises's quotation of Bergson's French, intuition (*sympathie*, or "feeling-knowing") is an awareness of the object's uniqueness that is otherwise inexpressible. Thus, when Mises says that concrete events cannot be "totally elucidated by logic," etc., he is expressing the premise that universal concepts cannot know concretes, thus his endorsement of the specific understanding as a form of intuition.

As such, intuition is a non-sensory, non-conceptual means of immediate awareness of reality, making it antithetical to reason and therefore mystical. Kant, in contrast, argues that intuition (*Anschauung*) is a type of perception (or experience), but one that is aware only of appearances, not the noumena of true reality, because it is filtered through the a priori lenses of space and time.[41] Which means reason and concepts can only give us appearances, leaving faith as the means of perceiving true reality. Hence, Kant's statement that he had to deny knowledge to make room for faith.[42]

In common usage today, intuition, or "intuitive insight," means the sudden grasp or understanding of a solution to a problem. The "insight" comes from our subconsciously stored knowledge related to the problem. A medical doctor, for example, confronted with ambiguous symptoms in the patient, gets a "feeling," "gut reaction," or intuition, that something more serious is going on. The intuition is the subconscious talking to the doctor.

A correct theory of concepts, however, begins with reality and the primacy-of-existence premise of Aristotle. It begins with the awareness of concretes through sense-perception. After the initial formation of concepts and later acquisition of a quantity of knowledge, the cognitive

---

[39] Mises, *Human Action*, 49–50.

[40] See above, p. 40, where I cite Jonathan Barnes who translates Aristotle's *nous* as *comprehension*, not intuition, of the universal in the individual.

[41] Kant, *Prolegomena*, esp. 30–33, 45–48. Immanuel Kant, *Critique of Pure Reason* trans. and ed. Paul Guyer and Allen W. Wood (Cambridge University Press, 1998), 185. First published in German in 1781, revised in 1787. This translation includes variations between the two German editions.

[42] Kant, preface to second edition, *Critique*, 117.

power that Mises is talking about is not a "specific understanding" or "intuition," but application, or conceptual identification to concretes, of previously formed universal concepts, a deductive process. Since our concepts refer to all concretes of a particular type, and may include what Rand metaphorically calls "mental concretes," or "mental entities," that is, already formed concepts, for example, of chairs, tables, and beds that we integrate into a new concept, an abstraction from abstractions called furniture, we live our daily lives applying previously learned knowledge to identify the specifics we confront.[43]

This, again, is how Sherlock Holmes deduced that Watson just came back from Afghanistan and our medical doctor deduces that our cough and runny nose are instances—symptoms—of a cold. Application, as I have said, is the mental process of identifying a this (concrete) as an instance of a that (concept). It is what historians do and what we all do every day.

Theory, History, and Measurement

Mises, in *Human Action,* writes, "Without theory . . . there is no comprehension of the reality of human action," and in *Epistemological Problems,* he states, "Concepts are always logically prior to the understanding of the individual, the unique, and the non-repeatable."[44] Only in this sense is Mises correct: as educated adults, we do need to have accurate theory to correctly identify the concretes of history, our own or the subject of the books a historian may be writing. This is the sense in which theory "precedes" history. The formation of theory, however, requires perceptual, then conceptual, awareness of human action, past and present, before the theory can be applied to write history. Aristotle and Rand are correct when they say all knowledge begins with perception. To call theory "a priori," unfortunately, is Platonist and sounds like we are condemned to looking at the shadows in Plato's cave. It introduces confusion into the discussion caused by acceptance of the primacy-of-consciousness premise.

Discussing the issue of measurement and quantitative analysis in economics, Mises states correctly that there exists "a regularity in the sequence and interdependence of market phenomena" but no "constant

---

[43] See Rand, *Objectivist Epistemology,* 21, 84, 156–57 for her use of the terms "mental concrete" and "mental entity."

[44] Mises, *Human Action,* 39, and *Epistemological Problems,* 1.

relations" or standard unit that can be used to make quantitatively exact measurements of human behavior. Which means no exact algebraic equations can be identified or used as in physics and chemistry.[45]

Human behavior is not deterministic because we have free will, which Mises does not explicitly endorse, though he does say that human action is teleological, which is another instance of his Aristotelianism. Because of the observed regularity that is present in human behavior, principles and laws of economics, such as the laws of marginal utility and supply and demand, measurements omitted, can be formulated and used to interpret current and past economic events and to make predictions of the future, albeit not mathematically precise predictions. This is Mises's rejection of the positivist demands for a mathematical economics.

Let me here elaborate a technical issue. The physical sciences measure the world using what are called interval and ratio scales, the former using equal intervals from one number to the next (1 to 2 is equal to 4 to 5) and the latter requiring equal intervals and a true zero point to produce valid ratios (for example, Kelvin temperature, but not Fahrenheit or Celsius, which are scaled at the interval level). In the human sciences, measurement is teleological using ordinal numbers and ordinal scales (A is better than or preferred to B and B to C). Relationships of means to ends are ordered or ranked because we do not have a unit that serves as a standard of measure in the way interval and ratio scales do. The intensity of preference in the intervals of two people preferring A to B may or may not be equal; we do not know.[46]

---

[45] Mises, *Human Action*, 1, 56.

[46] See S. S. Stevens, "On the Theory of Scales of Measurement," *Science* 103, no. 2684 (June 7, 1946), 677–80, https://www.jstor.org/stable/1671815, and Rand, *Objectivist Epistemology*, 32–34, 223–25. Note that ordinal numbers are quantitative in the sense that we can assign the numbers one, two and three, or rather, first, second, and third, for example, to our rankings. They just are not precise as in the interval or ratio levels of measurement. Self-report questionnaires, using, say, a five-point scale to measure concepts of consciousness, including value, are grossly approximate. Temperature, before the invention of thermometers, was also measured approximately, though blacksmiths knew their fires were sufficiently hot by the color of the heated ingots. Will scientists someday discover a standard unit by which to measure concepts of consciousness at the interval or ratio level? I do not know. See Kirkpatrick, *In Defense of Advertising*, 154–56, and John P. McCaskey, "History of 'Temperature': Maturation of a Measurement Concept," *Annals of Science* 77, no. 4 (2020): 399–444, https://doi.org/10.1080/00033790.2020.1817980.

The inability to formulate mathematically precise principles and equations is true of all human sciences, basic and applied, not to mention all the branches of philosophy. Positivism has corrupted our understanding of science, the actual essence of which is conceptualization, which means measurement omission even in the formulation of the equations of physical science.[47] Most so-called exact measurements in the human sciences, especially statistics, produce historical data, not universal, theoretical knowledge of cause-and-effect relationships.

In the first paragraph of the preface to the *Ultimate Foundation of Economic Science*, Mises writes: "This essay is not a contribution to philosophy. It is merely the exposition of certain ideas that attempts to deal with the theory of knowledge ought to take into full account." Mises in dealing with these issues, despite his use of some Kantian terminology, is taking steps toward a post-Kantian, modern Aristotelianism in economics—meaning, as Freud accomplished in psychology, upholding the active and reasoning, creative consciousness, guided by the tool of logic, as the means by which we know reality. Those who would advance economics further can build on Mises's work.

Those who would advance any science would be well-rewarded by reading Mises on epistemology. I include students of Ayn Rand, as they would not just learn Mises's economics or his approach to epistemology, but how his mind works to conceptualize science.

ON SUBJECTIVE VALUE

Let us move on to the concept of value, which underlies and is presupposed by economic demand. Mises is mostly correct when he states, "Value is not intrinsic, it is not in things. It is within us; it is the way in which man reacts to the conditions of his environment."[48] Values, and all concepts, according to Rand's theory, are within us. Essences and values are not "out there" in the thing; they are not metaphysical. That would be moderate realism. Using the tools, or mental processes, of our consciousness, we form concepts, including concepts of what is valuable. But values are objective in the *epistemological* sense—if they are a correct identification of what contributes to the flourishing of human life, one's own life in particular.

---

[47] See Kirkpatrick, *In Defense of Advertising*, 153–58, and Kirkpatrick, *Independent Judgment*, 72–81.

[48] Mises, *Human Action*, 96.

We can distinguish between object and subject in the sense that objects are "out there," while whatever goes on in our minds is subjective, but that is the different usage I mentioned and cautioned about in chapter 2.[49] Customer value judgments are subjective because their judgments come from inside their heads, but they may (or may not) be epistemologically objective—correctly identified as beneficial for or harmful to them. To put it a different way, the *objects* of our awareness are metaphysical, but our *means* of awareness, which is fallible, is epistemological.

Mises—and the positivists, and Mises's friend, Max Weber—talk about the necessity of all sciences being "value-free," which means "shoulds" and "oughts" do not belong in science. This derives from Kant who faced a dilemma between ethical values that come from the nature of the mind (but cut off from the requirements of human life) and free will that comes from the noumenal world but is unknowable.[50] Positivists "resolved" the dilemma by explicitly making all values subjective, meaning arbitrary.

Ayn Rand's Theory of Value

Ayn Rand's theory—a general theory, as I call it—can be widely applied. Rand defines value as "that which one acts to gain and/or keep" and applies it to all living organisms. The concept of value presupposes answers to the questions "to whom?" (or "what?" in the case of lower living species), and "for what purpose?" A value is whatever sustains and enhances the organism's life, though mistakes and misjudgments, in the case of human beings, can always occur.[51] A disvalue, perceived as a threat, is what humans (and all living organisms) avoid or attempt to destroy. Values and disvalues are species of the concept *fact*, based on how beneficial or harmful they are to life.[52] In the conceptual hier-

---

[49] See above, p. 51.

[50] The word "transcendental" in Kant's philosophy means beyond reason and experience. Since free will—and God and immortality—are in the noumenal world, Kant claims to know them through transcendence, not reason, which can only know phenomena.

[51] Ayn Rand, *"Objectivist Ethics,"* 5.

[52] This is why in *Independent Judgment*, 150 n24, I wrote that human value judgments are fact judgments. See Rand, *"Objectivist Ethics,"* 7–8, 15, for her answer to the alleged "is/ought" problem in ethics.

archy, the concept of value presupposes the concept of life. It is not possible, except by bad and mistaken metaphor, to talk about inanimate objects having values.[53]

Can we apply this definition to the economic concepts of value and utility, especially as used by the Austrian economists? When I was teaching, I would tell my students that the law of marginal utility can more clearly be understood as the law of marginal value. Was I correct?

The answer to Rand's second question, "for what purpose" as applied to all living organisms, is to sustain and enhance that organism's life. The amoeba, with its pseudopodia extended, takes in nourishment from the environment, the plant strives to get sunlight and nitrogen, and the family dog stares at us and gives an affectionate growl to communicate when it is time for dinner.[54]

For human beings, the concept of value is broad enough to cover the universal moral values of ethics, as well as the many non-moral, optional values of applied science and everyday life. While Rand uses the concept to develop her theory of ethics, identifying the life of a human, a rational animal, as the standard of morality, her concept applies to every action humans take in life, moral or immoral, healthy or unhealthy, law-abiding or criminal, productive or indolent, correct or mistaken. Human beings—each one of us as individuals, and because we possess free will—must choose what we judge will sustain and enhance our lives. We must declare something to be a value, then pursue it. Value is not "out there," intrinsic in the things that we like, because it is a relationship between us and the thing that we rate (evaluate) as beneficial. And, again, we may be mistaken in our judgment or choose what is harmful either to ourselves or to someone else.

To illustrate further the meaning and role of values in our lives, some of us may choose honesty as a way of life, others dishonesty or theft to acquire values. Some may build bridges for a living, while others drive trucks. Some may eat vanilla ice cream, while others prefer chocolate, and still others may not like ice cream at all.

Many values that we pursue—a specific food, or career, or romantic partner—are optional, that is, not universal, which means they do not have to be pursued by everyone (nor, obviously, would we want

---

[53] Rand, 7.

[54] The amoeba's pseudopodia is a favorite analogy of Freud's to human behavior, found in several places. We extend *the I*, he says, to objects (people and things) and bring them back to our libido to enhance our lives. Freud, *Outline*, 7, in *Penguin Freud Reader.*

everyone pursuing our romantic partner), whereas moral values are universal requirements for the sustenance and improvement of human life as a rational being. Moral values, the ends we are pursuing, and virtues, the means or actions taken to acquire and use the values, are broad abstractions, such as honesty, courage, and integrity, with the corresponding virtues, respectively, of telling the truth, acting against great odds or opposition, and remaining loyal to one's fundamental convictions. Moral values and virtues are not concretes, such as red meat, or plastic water bottles, or guns. I call this the fallacy of moralizing concretes.[55] Moral values and virtues are abstractions that each one of us must apply to the specifics of our lives.

Rand's concept of honesty is not Kant's dictum of never lying—always acting from duty, not inclination.[56] The principle of honesty correctly stated is to tell the truth, unless threatened with physical force or invasion of privacy, or when a blunt truth might unnecessarily hurt another person. Rand's other moral values (and virtues, in parentheses): reason, the most fundamental one (rationality, always following reason), independence (relying on one's own mind to perceive reality), purpose in life (productivity, pursuit of a rewarding career), justice (judging oneself and others correctly and, when appropriate, treating accordingly), and self-esteem (pride, Aristotle's crown of the virtues, upholding one's self and life as the highest value, never sacrificing oneself to others nor others to oneself).[57]

Judging others as moral or immoral is not an easy task, because psychological development from our early years plays a significant role in the way we behave in the present. The criminal personality, however, is decidedly immoral; such a person lies, cheats, and steals as a way of life and enjoys getting away with the forbidden.[58] Self-esteem and pride, it should be noted, each have two interrelated meanings. The moral one as stated above and the psychological one: our conviction of worthiness and mental competence to live (psychological self-esteem) and

---

[55] Not every issue, in other words, is a moral issue. It is how the person uses the concrete object, say, a gun, that makes the issue a moral one or not.

[56] Immanuel Kant, *Groundwork of the Metaphysics of Morals,* trans. Mary Gregor (Cambridge University Press, 1998), esp. sec. 1. First published in German in 1785.

[57] Rand, *"Objectivist Ethics,"* 19–22.

[58] See Samenow, *Inside the Criminal Mind*, esp. chap. 12. Yochelson and Samenow, *Criminal Personality*, esp. chap. 7, describe what they call the "nonarrestable criminal" who does not per se violate law, but under cover of family and job lies, mooches, cheats, manipulates, and intimidates.

the emotional consequence and expression of this conviction (pride). Psychological pride in oneself is a quiet confidence of accomplishment, not a brag or boast.

Consider again the meaning of value and utility in economics, taking value first. Austrian economists Carl Menger and Ludwig von Mises define value as the importance of something to us, Menger adding that it satisfies a need to maintain our lives and well-being. Mises goes on to emphasize that value means preferring one thing to another. Both consider values to be subjective.[59]

The words "importance" and "preference" are close synonyms of "value," making these definitions circular: "I value what is important to me and I value what I prefer." These economic concepts do not clarify the meaning of value. Menger's additions that values satisfy needs is closer to Rand's, but he maintains that our needs are changeable, hence making values subjective.

Rand's definition is general with a broad genus ("that"—anything) and differentia (which "one acts to gain and/or keep").[60] For human beings, it is the latter, our action, on the basis of our judgments, that actualizes those things outside of us to make them our values. Iron ore and oil, for example, before they were extracted and turned into steel and gasoline, did not have value for anyone—until entrepreneurs acted to turn them into economic goods.

Economic goods are things that have been turned into values that consumers desire and pursue, in contrast to "free goods," such as oxygen in the air we breathe. Oxygen is a value required for life, but we do not have to pay a price for it, that is, unless, for example, we are going deep sea diving. Then it becomes a highly valued economic good.[61]

---

[59] Carl Menger, *Principles of Economics*, trans James Dingwall and Bert F. Hoselitz (1976; repr. New York University Press, 1981 115, 121. First published in German in 1871. Mises, *Human Action*, 96. Cf. Hülsmann, Introduction to *Epistemological Problems*, xxxiv–xli.

[60] The genus of human values can also be stated as anything we judge as sustaining and enhancing our lives (keeping in mind that mistakes can be made and harmful, anti-life conduct can be willfully chosen.) The differentia remains that action is taken to acquire and use the judged object.

[61] Cf. Reisman, *Capitalism*, 40–41. Exchanges in an economic context are usually described as a mutual giving up of something we value less in order to receive in return something we value more. Buyers give up money for the more desired goods and sellers give up goods for the more desired money. In the non-economic context of personal and moral relationships, we assume a commitment to exchanging value for value as a matter of justice. I doubt that a good

The economic concept of "utility," as I also would say to my students, and not intending to be flippant, does not seem to be especially useful. Its meaning is "usefulness" or "serviceableness" to meet our needs, which again is a synonym of value. It predates the philosophy of utilitarianism but was popularized by the utilitarians who introduced errors into ethics and economics, not least the subjectivism, relativism, and collectivism of values.

The law of marginal utility can indeed be stated as the law of marginal value, meaning that the larger the quantity of something that is available to us as a value, the lower the economic value we will place on a single (or "marginal") unit. The law, of course, resolves the paradox of value of the classical economists who could not explain why diamonds are so much more valuable in economic terms, that is, higher priced, than water, the answer being the quantity of each that is available. Stranded in a desert, fully dehydrated, we would value a glass of water as high as our own life. But neither value—diamonds or water—is subjective. Values are contextual and determined by each individual in his or her situation, requiring answers to the questions "to whom?" and "for what purpose?" They are not metaphysically subjective.

In psychology, the theory of perceived risk is often explained as a *subjective* risk that we may experience. For example, when getting on an airplane, it means risk varies with each individual. One person needs alcoholic beverages before boarding; another enjoys walking on wings for a living. The risk perception is individualistic.

Thus, I submit, we could substitute the words "perceived" or "individualistic" for the word "subjective" that the Austrian economists use to modify "value" without losing accurate meaning of the concept. And perhaps that would clarify some of the confusion that comes up in philosophical discussions between objective and subjective value: values are not "out there" in the world, but they may be epistemologically objective, which means most food is epistemologically objective, while choice of what is good or valuable food will vary widely from person to person. Certain species of mushroom, however, will, or should, not be valued as food.

relationship would last if one person said to his or her spouse, "You get a lot of value from me, but I get less from you." See Rand, "Objectivist Ethics," 28–29, on the trader principle.

Rand's definition of value gives us a theory applicable to all living organisms, including the wide range and types of values held by human beings. All human action, to borrow Mises's words, is guided by a goal that becomes the standard by which we identify the steps or means—the intervening values between plan and end—required to achieve the goal. The goal is the ultimate end or value that we seek. This description of human action applies to ethics, economics, psychology, engineering, and all other concepts of method. Every action we take is motivated by our desire to acquire and use a specific value or to avoid a disvalue.

Felt Uneasiness

This last comment about motivation by chosen values differs from Mises who says that we are motivated by a "felt uneasiness" to seek those values. Sometimes we might be so motivated, for example when hungry and our stomach growls. Sometimes we act out of habit, such as eating lunch because it is time, not because we feel an "uneasiness" in our empty stomachs. Sometimes, for example, due to the pleasure experienced during a past orchestra performance, expected enjoyment becomes the motivator to attend another concert. Pleasure, if life-sustaining, is the consequence of acquiring a value. Our conscious mind—through free will—chooses and determines our values and disvalues, and what we will seek or avoid, by judging correctly or incorrectly what we believe to be good for us—enhancing our lives—or not. It is the power of our conscious minds that chooses and determines how we will act or not act.[62]

Mises cites John Locke and Gottfried Leibniz as sources of the concept of uneasiness. Locke offers an extensive discussion, associating the ideas of delight and uneasiness with pleasure and pain. The uneasiness of pain motivates us to seek relief, he writes, giving us delight or pleasure, which is essentially psychological hedonism, namely that we, by nature, pursue pleasure and avoid pain. The notion of "felt

[62] Pleasure and pain, as well as the emotions of joy and suffering, are consequences of acquiring, or failing to acquire, values. As such, they are indicators of life-enhancing or life-impeding actions. They are not primaries. See Rand, *"Objectivist Ethics,"* 23–25. Cf. Reardan, *Emotions*, 79–82. Aristotle identified pleasure, but not emotions, as an attribute of well-functioning life-activities. See *Nicomachean Ethics*, 1173b8–1176a28, in McKeon, *Basic Works*, 1096–1102. Plato said pleasure indicates life-enhancing activity by removing pain, not unlike Mises's notion that consumption removes uneasiness. *Philebus*, 31d, 32e. Aristotle, as I have stated before, was an improvement on Plato.

uneasiness," unfortunately, can easily justify criminal behavior in the sense that if you make me feel uneasy, I can harm you to give me relief from the uneasiness.[63]

Here are two intriguing quotations from Mises about values: "By means of its subjectivism, the modern theory becomes objective science" And: "it is in this subjectivism that the objectivity of our science lies."[64] Mises is stating that the theory of marginal utility answers the water-diamond paradox of the classical economists who had assumed values are "out there" intrinsic to reality. If "subjective" is understood as explained above and values can be viewed as epistemologically objective, then Mises is correct.

I prefer to call our values, market values in particular, "psychological," which makes them epistemological, neither metaphysical nor subjective and arbitrary. Rewording Mises's second quotation above, I would say the following: "It is in this recognition that values are psychological, not intrinsic or metaphysical, that the validation of economics lies."

Ayn Rand distinguishes between market value, or price, as socially objective value, "the sum of the individual judgments of all the men involved in trade at a given time." And philosophically objective value, "as value estimated from the standpoint of the best possible to man." For example, the airplane, in terms of speed and convenience of travel, is philosophically more valuable than the automobile, and the automobile is more valuable than the horse and buggy.[65] Understanding

---

[63] Mises, *Human Action*, 13. John Locke, *An Essay Concerning Human Understanding*, abbr. and ed. A. D. Woozley (1690; repr. Meridian, 1974), esp. bk. 2, chap. 7. Freud's pleasure principle made the issue more explicit as biological drives that urge us to act when they are not being satisfied, though the reality principle of *the I* can overrule the drives. This enables Freud to escape the charge of psychological hedonism. The utilitarians, in their attempt to come up with an objective ethics, made hedonism social and explicitly collectivistic, by making the "greater good" or "greatest happiness of the greatest number" the standard of value, implying, if not explicitly advocating, that the minority must be sacrificed to the majority. See Mises, *Theory and History*, 67–68, on Jeremy Bentham's failure to distinguish between "despotism and liberty." Mises insists that praxeology, focusing on the individual's pursuit of means in relation to ends, avoids utilitarian despotism.

[64] Mises, *Human Action*, 21, and Mises, *Epistemological Problems*, 192.

[65] Rand, "What Is Capitalism?" 16–17. First published in *The Objectivist Newsletter* 4, nos. 11–12 (1965). See also my discussion in *In Defense of Advertising*, 95–99, 175–78. "Socially objective" refers to all market participants bidding to buy and offering to sell a specific good. What determines the bid and ask prices? This is where the economist steps in to discuss supply and demand, marginal utility,

this distinction between the social and philosophical explains why philosophically objective values are not much involved in economics, which means no such philosophically valuable good should or ought to be mandated or forbidden by law, or priced a certain way.[66]

The criteria for bringing philosophical, especially moral, values into an economic discussion are the beneficial or harmful consequences of a specific policy to the citizens' lives. Thus, one can say that price controls are both immoral and economically impractical—immoral because of the harm to human life caused by injecting coercion into voluntary trade along with the consequent shortages that result, and impractical because wealth in the aggregate is decreased, not increased. It is only increased, if at all, for certain privileged groups.

Yes, there are those today who say that all issues are moral and political proclaiming "consumers should or ought not to buy that product" or "that bridge should not be built in that location."[67] In a free society where individual rights, especially property rights, are respected, shoulds and oughts would be minimal. In an interventionist economy, however, should and ought can readily be applied to the government, such as, the government *should not* (as a moral issue) impose price and wage controls on anyone. Applied sciences, in providing guidance to the achievement of particular goals, presuppose that the actions to be taken are moral and legal. This is the assumed context because the aim of applied science, all sciences, is to improve human life.[68]

Thus, in policy discussions, one must keep three value issues clearly separated: the moral, the legal, and the practical. Is the policy right or wrong morally, that is, beneficial or harmful to human life? What does the law say on the issue, and should the policy be legal or illegal? And will it work, that is, achieve its goal?

Values in all three areas are the motivators. "Felt uneasiness" may or may not be relevant.

---

and time preference. Market prices, in other words, are not arbitrary, uncaused, or inexplicable. They are socially determined and can be explained, which makes them, using Rand's term, epistemologically objective.

[66] Nor *should* the Pennsylvania Amish, for example, be coerced to give up their horses and buggies.

[67] In both examples, these people are moralizing concretes.

[68] To broaden the shoulds or oughts as applied to government: stay out of our professional and personal lives and arrest those who use or threaten to use physical force against us.

THE CONCEPT AND DEFINITION OF ECONOMICS

The reason I have been saying economics is a practical science is that its aim is to establish principles of how peaceably to increase wealth for everyone. Economics is the science of the production of wealth under a division of labor, not, as many textbooks insist, the allocation of scarce resources, nor, as Mises often puts it in his praxeological terms, the determination of means by which to achieve ends, which means the study of human choices.[69] Psychology, however, studies choices, as well as means and ends in the pursuit of values, and economics as the production of wealth requires allocation of scarce resources, whether performed by a socialist state or by the individual participants in the market.

The definitions of economics used today, including "catallactics," the term adopted by Mises, are nonessential. They do not explain or identify the causes that create material wealth. They assume the existence of wealth "somehow" created and then allocated or chosen by market participants. Production of wealth is the broad abstraction; allocation is derivative. Means, ends, and choice are broader concepts deriving from psychology and ethics. Scarcity is an insufficient or non-existent supply of whatever we need or want to satisfy our physical and mental needs. It derives from biology and confronts all living organisms.

The original meaning of the Greek word *oikonomia* is management of the household. The term "economy," according to the unabridged dictionary, refers to a system of meeting needs that is managed by individuals, households, businesses, governments, or even species.[70] When "political" is added as modifier, the concept implies management by the state, as it was understood in the age of mercantilism, which today could also mean socialism. The classical economists defined political economy as the production of wealth for both citizens and sovereign, but with state regulation diminished.[71] "Market" is the better modifier of economy, with emphasis on individual participants who manage

---

[69] Reisman, *Capitalism*, 15–16. Reisman includes "division of labor" to emphasize the importance of labor specialization in enabling the production of wealth. Cf. Mises, *Human Action*, 157–58.

[70] *Merriam-Webster's Unabridged Dictionary*, s.v. "system," accessed June 9, 2025, https://unabridged.merriam-webster.com/unabridged/system. "Economy" is in the synonym discussion for the word "system."

[71] Adam Smith, *An Inquiry in the Nature and Causes of the Wealth of Nations*, 2 vols. in one, ed. Edwin Cannan (University Chicago Press, 1976), vol. 1, 449.

their own affairs in the process of producing and consuming material values. Economics has become the settled term in use for many years, but its definition needs clarification.

I am arguing here that the essential distinguishing characteristic of the concept "economics" must be production, that is, the science (genus) of the process of creating and delivering material values—goods and services—to the market to meet the participants' needs and wants (differentia).[72] Focus on means and ends, choices, and allocation is not fundamental. As former Mises student, George Reisman, writes, "To claim that economics is on this account a science of human choices rather than of wealth is to confuse an aspect of the science with its totality."[73] It is nonessential by citing an effect instead of its cause. Focusing on means required to achieve ends is vague and can be understood as what engineers do, thus opening advocates of capitalism to the charge of endorsing "social" engineering. Means and ends, choices, and allocation are all involved in wealth productivity, but they are not the essence of economics.

"Catallactics," a third term not coined by Mises but used by him, means the science of exchange ratios and prices, or, more generally, the science of the market economy.[74] The word was apparently coined as a replacement for the nineteenth-century confusions of political economy.[75] Mises uses it as his replacement for economics. Unfortunately, the question must be asked: how valuable is it to have another coined word, cumbersome as it seems, when "economics" is not just widely used today, but, relatively speaking, understood at some level as dealing with production and trade.

Just as the capacity to reason is the essence of homo sapiens that explains and causes the ability to build skyscrapers, so also is production the essence of economics that explains and causes the generation of exchange ratios and prices. Robinson Crusoe must catch fish before he can take them to market to trade for other goods. The conceptual hierarchy is production first, then prices. The concept "production," after

---

[72] "Needs and wants," unfortunately, are seldom defined or differentiated. Needs are objective requirements to sustain and improve our lives. Wants are optional tastes and preferences.

[73] Reisman, *Capitalism*, 41–42. Chapters 1 and 2 of *Capitalism* provide extensive discussions of the role of wealth in human life and the proper conceptualization of economics.

[74] Praxeology and thymology are the other two terms.

[75] Mises, *Human Action*, 3.

all, as Mises points out, is a broad concept that includes advertising and distribution, which means "allocation" is derivative from production.[76] Prices then become the connecting link, or "communicator," as Hayek says, between producers and consumers, but production remains the fundamental essential distinguishing characteristic of economics that summarizes and condenses the content of the concept.[77]

Discussions of means and ends, choices, allocations, and scarcity are contained in the "file folder" labeled "economics," but they do not summarize or condense all or most of that folder's content.

THE BASIC AND APPLIED SCIENCE OF MISES

All sciences have their basic and applied components. Some scientists spend most of their time in the basics, seeking to push back the frontiers of their science, identifying new, revised, or extended meanings of the science's fundamental concepts, its theory. Others work in applied areas, providing guidance, recommendations for practice, and may also conduct historical studies of the field. Many scientists spend their time alternating back and forth between the basic and applied versions with criticism of previous work often occurring in both areas. Economists tend to spend much of their time on policy analysis and recommendations to improve economic life in society, as well as on economic history. Ludwig von Mises excelled at both basic and applied economics, producing also a large literature of criticism and economic history.

Unlike chapter 4 in which I spent a great deal of time, in separate sections, summarizing the essentials of Freud's basic and applied science, the present chapter, because most of it thus far has been in discussion and criticism of Mises's epistemology, especially his concepts of praxeology, thymology, and catallactics, I will summarize Mises's basic and applied science more briefly than I did with Freud. Also, because the science of economics is over a hundred years older and therefore more

---

[76] The significance of understanding this conceptual hierarchy explains, at least in part, what led George Reisman to identify profits as the original and primary form of income, with costs, labor costs in particular, deducted from profits, not the other way around as Marx and, unfortunately, the classical economists, assumed. Reisman, *Capitalism*, 473–85. Cf. my paper "Reisman's Net Consumption, Net Investment Theory of Aggregate Profit: Exposition and Comparison to Mises and Rothbard," *The American Journal of Economics and Sociology* 63, no. 3 (2004), 627–46, https://www.jstor.org/stable/3488146.

[77] Mises, *Human Action*, 316–19. Hayek, "Use of Knowledge in Society," 86–89.

developed than the modern psychology of Freud, much of Mises's work consists in revising and refining earlier-formed concepts, though he does also make wholly original identifications.

Here are highlights, presented as a kind of annotated bibliography of his most significant works.[78] The discussion does not exhaust Mises's accomplishments. I have left out some works of collected essays and all posthumously published lectures and collections. The summary, I hope, provides a taste of Mises's method of thinking and writing, namely conceptualization, and, especially, his approach to analyzing, identifying, and presenting a non-positivist science.

Major Works

The Theory of Money and Credit, 1912. Drawing on the work of earlier Austrian economists, especially Menger and Böhm-Bawerk, Mises revised and refined the concept of money, tracing its value, its purchasing power, back to the time when gold and silver were used as commodities for productive purposes. This eventually became known as his concept of the regression theorem, original to him, that denies an infinite regress in the determination of money's value. In this process, he integrated the laws of supply and demand and marginal utility with the theory of money, quashing the then prevalent (and current) dichotomy between so-called micro- and macroeconomics.[79]

Mises also revised the classical economists' concepts of inflation and the business cycle, pointing out, by endorsing the Cantillon effect, that increases in the quantity of money, created almost entirely by the government and its central bank, spreads and redistributes wealth

---

[78] Not to be confused with the encyclopedic bibliography, compiled by Bettina Bien Greaves and Robert W. McGee, *Mises: An Annotated Bibliography* (Foundation for Economic Education, 1993). Available, along with a second volume, in electronic form, at mises.org. These bibliographies include not just works by Mises but also articles and books written about Mises.

[79] Ludwig von Mises, *The Theory of Money and Credit*, trans. H. E. Batson (1934; repr. Liberty Classics, 1981), esp. chap. 1 and 2. First published in German in 1912, a fourth part on monetary reconstruction added to the English edition in 1953. Similar to the challenges of translating Freud, see H. E. Batson's two-page discussion of turning Mises's German into English, especially the word *Umlaufsmittel*, which became "fiduciary media" in the text, but "credit" in the title of the work, 525–26. See Hülsmann, *Last Knight*, 217n7, for his discussion of the same concept and its translation. The title of the book, he states convincingly, should have been "The Theory of Money and Fiduciary Media." Hülsmann's biography of Mises, I must add, provides a wealth of historical context for Mises's life and work.

unevenly throughout the economy as a type of taxation, and that the business cycle of booms is caused by credit expansion that produces malinvestment followed by bankruptcies, unemployment, and financial contraction—in other words, busts. Mises at this time recognized that values are ordinal rankings, leading him to assert the impossibility of measuring values on an exact (interval or ratio) scale as in the physical sciences.[80]

Economic Calculation in the Socialist Commonwealth, 1920. This forty-eight-page article is a masterful application of free-market principles to socialism and is far more significant than its length suggests. The article identifies why socialism must inevitably fail, because the bureaucrats running a socialist economy cannot perform economic calculations.[81] Socialism fails, Mises wrote, because there is no private property to enable exchanges from one owner to another, only the socialist state that owns everything. In effect, socialism is a large household economy in which there are no money prices or exchanges. As a result, planning for production and distribution, especially for the factors of production, cannot be accomplished without calculation in money, of costs, sales revenue, and profits. Socialism also lacks an "intellectual division of labor" required to make the myriad decisions throughout the system to satisfy the needs of all participants in the market. Socialism requires omniscience to make those decisions and is therefore an attempt to manage the affairs of millions of people as if they were members of a single primitive household. "Socialism," Mises concludes, "is the abolition of rational economy."[82]

---

[80] Mises, *Theory of Money and Credit*, esp. chap. 5, sec. 5; chap. 6; and chap. 7, sec. 3. Mises also stated that money is not a measure of prices or value. Price is the quantity of money we are willing to spend for the product. Our evaluation of the good, which varies from person to person, determines how much money we are willing to part with for the good. Chap. 2. Cf. the concise presentations of Mises's theory of money in Rothbard, *Essential Von Mises*, chap. 1–3, and Hülsmann, *Last Knight*, chap. 6.

[81] In Ludwig von Mises, *Memoirs*, trans. Arlene Oost-Zinner (Mises Institute, 2009), 17, Mises writes, "It was the study of political economy . . . that led me to an understanding of the true nature of interventionism." This statement refers to his university years, probably 1907–08, when he was beginning to doubt the validity of interventionism. His study of political economy also led to his understanding and refutation of socialism. The *Memoirs* were written shortly after Mises's arrival in the United States in 1940. They were not published until 1978, posthumously in both German and English.

[82] Ludwig von Mises, *Economic Calculation in the Socialist Commonwealth*, trans. S. Adler (Mises Institute, 1990), 15, 23. First published in German in 1920 and translated into English in 1935.

Socialism: An Economic and Sociological Analysis, 1922. In this work, the second of Mises's several magna opera (*The Theory of Money and Credit* being the first), the "abolition of rational economy" became *destructionism*, because socialism is unable to create anything; it can only destroy.[83] The brevity of my comments about this exhaustive criticism of all forms of socialism known in his time does not decrease my respect for the work. Many of those forms of socialism are still preached today. In this work, Mises broadens Ricardo's law of comparative costs (advantage), renaming it (in *Human Action*) the law of association, because free-market capitalism creates, through peace and trade in the division of labor, incentives to increase personal and professional cooperation regardless of skill or ability, thereby leading to increases in everyone's standard of living.[84]

The 1951 edition of the English translation of *Socialism* includes the epilogue "Planned Chaos," originally published in 1947, a historical essay on the rise of Soviet communism. In this essay, Mises distinguishes communism from fascism and Nazism, the latter two systems identified as "socialism of the German pattern."[85]

Human Action, 1949. English version of *Nationalökonomie*, 1940. Mises's greatest work, his magnum opus, of both basic and applied science is *Human Action*, to which I have made ample reference in this chapter. It is a comprehensive treatise on economics and includes much of his work on the science's theoretical—epistemological—foundations. It also exhaustively covers the fundamental concepts and principles of economics, in revised and refined form, including both what earlier economists had written, as well as the concepts and principles of his own earlier work. Topics (a highly abbreviated list) include: human cooperation; money, credit, and banking; marginal utility, time preference, value, and prices; the market, division of labor, and role of the entrepreneur; interest and profit; capital, labor, and wages; monopoly;

---

[83] Ludwig von Mises, *Socialism: An Economic and Sociological Analysis*, trans. J. Kahane (1936; repr. Yale University Press, 1951), pt. 5. First published in German in 1922.

[84] Mises, *Socialism*, 292–95; *Human Action*, 158–63.

[85] Mises, *Socialism*, 533. *Planned Chaos* (1947; repr. Mises Institute, 2014) is available as a monograph and includes this powerful identification, chap. 6: "The real significance of the Lenin revolution is to be seen in the fact that it was the bursting forth of the principle of unrestricted violence and oppression. It was the negation of all the political ideals that had for three thousand years guided the evolution of Western civilization."

and the business cycle. Interspersed throughout the work are discussions of economic history, applications of essential economic concepts to the past and present. The last two hundred pages of the book focus on the application of free market principles to his devastating critiques of interventionism and socialism.

Shorter Works

Nation, State, and Economy, 1919. "Contribution to the Politics and History of Our Time" is the subtitle, written in the aftermath of the first world war. Originally titled "Imperialism," Mises analyzes the causes of the war, the meaning of imperialism and statism, nationalism (two concepts, liberal and imperialistic), the right of self-determination (with emphasis on freedom for minorities), and socialism. The book is an application of free-market principles to these concepts.[86]

Liberalism in the Classical Tradition, 1927. Mises in this work clarifies the essential meaning of classical liberalism, with emphasis on private property, which is what enables us to enjoy freedom, peace, and prosperity. Free-market economics and foreign policy are also covered, and the book includes a discussion of democracy and the significance of voting as requirement for a peaceful succession of leadership.[87]

Critique of Interventionism, 1929. A collection of six essays from the earlier 1920s, this is the third of three works on social cooperation—*Socialism* and *Liberalism* being the first two. His method of analysis, Mises points out, is to analyze the means and ends proposed by interventionists and socialists and demonstrate that their ends cannot be attained. Interventionism unchecked leads to socialism and socialism destroys itself. "Capitalism," on the other hand, he writes, "does not destroy itself."[88]

---

[86] Ludwig von Mises, *Nation, State, and Economy: Contributions to the Politics and History of Our Time*, trans. Leland B. Yeager (New York University Press, 1983). First published in German in 1919.

[87] Ludwig von Mises, *Liberalism in the Classical Tradition*, 3rd ed., trans. Ralph Raico (Foundation for Economic Education and Cobden Press, 1985), 39–42. First published in German in 1927.

[88] Ludwig von Mises, *Critique of Interventionism*, trans. Hans F. Sennholz (1976; repr. Mises Institute, 2011). First published in German in 1929. Mises's quote and comment about the three works on social cooperation are in *Memoirs*, 97–98. On page 98 of the memoir, Mises writes this sad comment, shortly after his arrival in the early 1940s in the United States: "I set out to be a reformer, but only became a historian of decline."

Epistemological Problems in Economics, 1933. A collection of essays going back to 1928. It is also his first important work on epistemology, though the literal translation of the German title is "Basic [or Fundamental] Problems of Political Economy." Issues discussed: the science of human action, sociology (praxeology) and history, conception and understanding, subjective value (two chapters), and the difference between psychology and economics. The distinction between theory and history is prominent throughout.

Bureaucracy, 1944. In this short book, and first work written in English, Mises identified the nature and essence of bureaucracy, disagreeing with the commonly accepted view of Max Weber that is still taught today in business schools, namely that any large organization, government or business, that is hierarchically structured is a bureaucracy. Mises turned bureaucracy into an adjective modifying management. That is, *bureaucratic* management, with a budget, laws, and rules, is how governments manage their affairs, whereas *profit* (or business) management, using profit through customer satisfaction as the yardstick, is how profit-making businesses manage their affairs. To the extent that a business is regulated, having to comply with government-imposed laws and rules, to that extent it will be bureaucratic, sometimes looking and acting like a government bureau. Socialist dictatorships, Mises identified, operate entirely by bureaucratic rules.[89]

Omnipotent Government, 1944. The subtitle of this book states its essence: "The Rise of the Total State and Total War." It is about Germany, the collapse of German liberalism and rise of Nazism, and the future of civilization. Some issues from *Nation, State, and Economy* are reviewed and updated, such as statism and nationalism. Totalitarian rulers, Mises writes in the preface, claim to be "wiser and loftier than their subjects and . . . therefore know better what benefits those ruled than they themselves." Mises makes it clear that Nazism and fascism are not related to capitalism.[90]

Anti-Capitalistic Mentality, 1956. This book is about the envy and resentment people feel toward capitalism. The free market requires merit and accomplishment. As a result, it "de-proletarianizes the 'common man'" elevating such persons to the level of a "bourgeois." Those who have not accomplished or earned as much

[89] Ludwig von Mises, *Bureaucracy* (1944; repr. Mises Institute, 2008).

[90] Ludwig von Mises, *Omnipotent Government* (1944; repr. Mises Institute, 2010).

as their colleagues, as the pattern goes, resent the more successful and turn their hatred toward capitalism. Mises discusses a variety of such resenters and enviers.[91]

Theory and History, 1957. Subtitle: "An Interpretation of Social and Economic Evolution." The thrust of the work is a further presentation of Mises's point that sound history requires a sound theory to interpret the events of the past and how those events brought us to the present. Or, as biographer Hülsmann writes, *Theory and History* is "a sweeping epistemological vindication of the case for liberty and capitalism." And it is, despite my criticisms of Mises's epistemology presented earlier in this chapter. The work covers values and value judgments, determinism and causality, polemics against attacks on capitalism, and offers comments on several philosophies of history. Roy Childs, using words from Ayn Rand's definition of art, as well as familiar words from Mises, defines history as a "selective recreation of the events of the past, according to a historian's premises regarding what is important and his judgment concerning the nature of causality in human action."[92] Theory, in other words, determines one's view of history.[93]

Ultimate Foundation of Economic Science, 1962. Subtitled "An Essay on Method," this short work is Mises's last book, published when he was eighty-one years old. Highly concise, it is a restatement of the epistemological foundations of economics. He spares nothing for the "panphysicalism" of positivism, stating in his last sentence of the work, that positivism is "knowingly and intentionally undermining the intellectual foundations of Western civilization."[94]

MISES THE SCIENTIST AND CAPITALISM

Mises's method of doing science, as it was for Freud and Aristotle, was conceptualization, the method of forming and applying concepts. At times, Mises refers to his (non-quantitative) method as logical reasoning.

---

[91] Ludwig von Mises, *The Anti-Capitalistic Mentality* (1956; repr. Mises Institute, 2008). The quotation is on page 1.

[92] Hülsmann, *Last Knight*, 950. Roy A. Childs, Jr., "Big Business and the Rise of American Statism," in *Liberty Against Power: Essays by Roy A. Childs., Jr.*, ed. Joan Kennedy Taylor (Fox & Wilkes, 1994), 15–47.

[93] And, broadening the notion, I might add that whatever concepts and principles we hold in our minds, right or wrong, determines our understanding of reality.

[94] Mises, *Ultimate Foundation*, 132. Panphysicalism refers to the positivist premise that all science must follow the methods of physics, effectively reducing every science to materialism.

This is an accurate description of the methodology of all sciences, the human sciences in particular, since the latter do not use or require the quantitative techniques of the physical sciences. Logical reasoning is the mental process of identifying and stating the necessary steps, or arguments (premises), to arrive at a conclusion. The conclusion may be the formation and definition of a concept or the concept's application to identify a concrete, or it may be a chain of arguments to establish and validate a theory. It is the process of demonstrating the validity of numerous conclusions derived from numerous premises or arguments.

When Mises in his *Memoirs* refers to his use of the "step-by-step" method, he was describing logical reasoning.[95] It was this method that led Mises as a young student to recognize that interventionism must fail to achieve its stated goals. In later years, he recognized the same for socialism. Logical reasoning is what led Mises to identify laissez-faire capitalism as the only valid system that can ensure peace and prosperity.

[95] Mises, *Memoirs*, 45–47.

III. Final Thoughts

6

The Essence of Science, Revisited

The true form or definition of heat (of heat as a universal notion, not relative merely to sense) is, in a nutshell, as follows: heat is an expansive motion which is checked and struggling through the particles.

—Francis Bacon[1]

Bacon realized that the aim of experimental science is the refinement of our ideas about the natures or essences of the substances, properties and processes we find in the natural world. Typical scientific questions would be "What is colour?" "What is liquefaction?" "What is heat?" In answering such questions we would have to formulate definitions of the nature of these things, processes, properties.

—Rom Harré[2]

EVERY SCIENCE IDENTIFIES KEY CONCEPTS and principles to explain cause and effect relationships in that science's specialized area of existence, then integrates those findings into a body of knowledge.

Applied sciences, relying on the knowledge of the fundamental ones they rest on, identify new concepts and principles in their specific domain to guide human choices and actions in the achievement of goals.

---

[1] Francis Bacon, *The New Organon*, ed. Lisa Jardine and Michael Silverthorne (Cambridge University Press, 2000), bk. 2, aph. 20. Italics omitted. First published in Latin in 1620.

[2] Harré, *Great Experiments*, 13.

The ultimate aim of all sciences, basic and applied, is to improve human life. That is, the aim of science is not to perform "mental gymnastics," to borrow Mises's words. Science describes reality and aids humans to choose effective, desired ends and to act efficiently to reach them.

THINKING IN DEFINITIONS

Thinking in definitions, principles, and, especially, ranges of measurement are skills exhibited by scientists, though some may not be aware of the talent.

Definitions and principles are what constitute our knowledge. In fact, all properly defined concepts are principles. For example, the concept "human being," stated as a principle, says humans are the animal that possesses the capacity to reason, and, as a clarifying addendum, exercises that capacity in degrees varying from person to person. The genus links the concept to the wider category from which its referents are differentiated, the higher animals that possess a consciousness, and the differentia states the essential distinguishing characteristic.

The definition, to repeat from chapter 2, is not a tautology, equivalent to the concept; it is a summary and condensation of all our accumulated knowledge of the entity "human being," the accumulated knowledge being the "file folder" of information about humans that varies from child to adult and from adult to adult.[3] The definition in this example focuses on an entity. Other concepts and their definitions identify attributes abstracted from the entities that exhibit them, such as color and heat, to use Francis Bacon's examples, or identify processes that entities enact, such as liquefaction.

Thinking in ranges of measurement does not mean thinking in Newtonian algebraic equations, though it may for some concepts. It refers to our awareness when forming concepts of the "stopping points," or boundaries, of distinguishing characteristics that limit the concept to a certain range of measurements, with their specifics omitted in forming the concept.[4] Tables, for example, are not a mile long, nor are they only one inch long. Bridges are flat with supports and hold objects

---

[3] See above, p. 46.

[4] "Conceptual boundaries" or "boundaries of inclusion" is how McCaskey, "History of Temperature," 441–43, characterizes ranges of measurement. Recall that these measurements can be approximate in the formation of concepts, as they were before the discovery of light waves for color and the invention of thermometers for temperature. An important function of a definition, says Kelley, *Art of Reasoning*, 32–33, is to "clarify the *boundaries* of a concept." Kelley's italics.

called automobiles, but bridges are not tables. A dollhouse's toy table may be flat with supports and hold tiny teacups and saucers, but it is not the type of table at which adults eat dinner. The concept "human" is based on the range of measurements that describes our distinctive type of mental processing and distinguishes it from that of the higher animals, the lower and upper ends of the continuum being the point at which all humans with a normal brain can use reason to form and apply concepts, whereas the higher animals cannot.

To elaborate, using McCaskey's examples from chapters 1 and 3, the boundaries of cholera were narrowed to an "organic poison," after dismissing season of year and foul water as causes.[5] Robert Koch then further narrowed the cause and differentia to the bacterium he called the comma bacillus. When Charles Wells, after many experiments, narrowed the boundaries of dew to droplets made in accordance with the laws of heat, particularly liquefaction, distinguishing it in particular from guttation, he recognized that condensation from vapor in the air was the cause and differentia. In both cases, the range of measurements of the respective differentia, comma bacillus and water condensation, were sufficiently limited to establish the concepts as universal. Lord Kelvin concluded, after two hundred years of imprecise and often inaccurate tidal measurements, that Newton had been right, namely that tides are and result from gravitational pull of the sun and moon, measurements omitted.[6]

In answering Bacon's questions in this chapter's epigraph, Harré points out that we must "formulate definitions of the nature of these things [entities], processes, properties [attributes]." Definitions as principles are *descriptive*, and can readily be converted into *prescriptive* principles, depending on what goal we are seeking. For example, humans possess the capacity to reason, thus to achieve our chosen goals we must (should, ought) exercise that capacity—to the fullest degree we are capable of.

---

[5] See above, pp. 20 and 63.

[6] See McCaskey, "Induction in the Socratic Tradition," 183–86. In chapter 1 of the present work, above, p. 21, I point out that the abundance of tidal measurements at various ports was applied science, valuable to the sea captain as relevant historical data to use for prediction, but not essential to the theory that identifies what explains and causes tides. Narrowing the various types of rising and falling water near shore to the pull of gravity, after omitting such non-celestial factors as temperature variations and industrial releases, was the first step in identifying what was universal.

Options in the formation of concepts, as Ayn Rand points out, mean that different concepts may use different "stopping points" in the continuum. Hence, the English word "man" can be a synonym of human being or, using the range of measurements describing male and female anatomy, can mean a male person. The Yiddish word "mensch," now an accepted part of the English language, means a rational, volitional human being who acts with integrity and honor. Options in concept formation are what make translation from one language to another challenging, because the range of measurements used in the concept's formation differ from language to language. Keeping in mind the continuums that constitute our concepts enables us to maintain contact with reality, the reality of the different-but-similar referents we are discussing.[7]

Thinking in well-constructed definitions also enables us quickly to bring to mind—to conscious awareness—the knowledge of what we want to think about or express. This applies especially to abstractions from abstractions, more so than to the perceived concretes of our everyday world.

A final note in this section on definitions and thinking in definitions is that a favorite vehicle of argumentation by rationalists and positivists, unfortunately, is to make grossly unrealistic assumptions and then announce, "Let's assume, *by definition,* that dogs can fly (or that we are brains in a vat) and see where that takes us." If, by definition, they mean, and they usually do, arbitrary, disconnected floating abstractions, which violate all the rules of good definition, knowledge can be anything they want it to be, which, to say it one more time, is where we are today as consequence of Kant's philosophy and his positivist followers.[8]

---

[7] Cf. Freud's caution not to make sharp divisions or lines as on a map, between *the I, upper-I,* and *it* when discussing the continuum (or stream) of consciousness, *New Introductory Lectures,* 99.

[8] Flying dogs (and maybe brains in a vat) may make interesting fiction, but they do not constitute science. In economics, the doctrine of pure and perfect competition does not even make good fiction—"perfect information," that is, omniscience, just being one of its ridiculous assumptions. Kirkpatrick, *In Defense of Advertising,* 118–21. A concept, to emphasize, is not "true by definition." A concept, and its definition, is "true by reality," that is, true by our having correctly identified its referents in reality and having summarized and condensed the file folder's content for ease of use. Concepts are not "ampliative," a Scholastic word McCaskey uses. They are open-ended in the sense that they apply to all referents of the same kind, past, present, and future—because they are classifications and cross-classifications (in our minds) of reality.

LAWS AND PRINCIPLES

Where do laws and principles fit into this discussion? What exactly are they? Laws are principles but principles are not necessarily laws. Both are universal. "Principle" is the older term.

Slightly adapting a definition from the *Oxford English Dictionary*, principles are fundamental truths on which others depend that provide causal explanations and guidance to human action.[9] A principle as causal explanation on which others depend is Aristotle's *archē* or first principle, the foundation of a chain of reasoning, which means that some principles in the conceptual chain will be more fundamental than, or presupposed by, others. This chain of reasoning is analogous to the chain of an abstraction from abstractions of a single concept. The Latin word *principium* is a translation of Aristotle's *archē*. The *OED* defines principles as guides to action in a separate entry.

Here are two examples of principles and their derivative implications: boiling water from chapters 2 and 3 and lying from chapter 5.[10] Water boils at 212 degrees Fahrenheit at sea level, but at lower temperatures at higher elevations or at higher temperatures if the water contains impurities, such as salt. This is a causal explanation and an implicit guidance to adjust cooking recipes in high altitude and for salted or sugared water. Never lie in human relationships, unless threatened with initiated coercion, uninvited invasion of privacy, or when a blunt truth may unnecessarily hurt the other person. This is a guide to human action, a moral one at that, and an implied causal explanation of what is required for benevolent cooperation and friendship.

Principles are combinations of concepts that make connections—between the genus of an entity and its essential distinguishing attribute, the motions or actions of the entity with its attributes, or the relationship between the attribute(s) of one entity and that of another. In the less clear but not unusable parlance of contemporary (positivist-influenced) science, principles connect one variable to another. The connection may be extremely precise using an algebraic equation consisting of constants and variables or it may lack that quantitative precision at the level of interval or ratio scales.

9 *Oxford English Dictionary*, "principle (n.), sense II.3.a," March 2025, https://doi.org/10.1093/OED/1796119223.

10 See above, pp. 54, 63, and 135.

The law of supply and demand, for example, only recognizes prices as varying in an inverse relationship with supply and a direct relationship with demand, although it can be expressed algebraically as $p = d/s$.[11] Any results produced by the formula are approximate because human beings with free will are present in both the numerator and denominator. This means sellers and buyers can change their minds and desires at any time, thus preventing a standard unit from being identified to make precise measurements. All terms of any equation that may be devised in the human sciences use concepts of consciousness, which means mental processes that can only be measured approximately at the ordinal level.[12]

Laws do the same thing that principles do, though with less emphasis on guidance. The term arose in the sixteenth and seventeenth centuries of the scientific revolution, usually as "laws of nature" and often only associated with the physical sciences, including those sciences' notions of experimental method and quantitative analysis. Economists, especially the Austrians, have adopted the concept of law, for example, of supply and demand and marginal utility, as a principle that makes broad and sweeping generalizations, which means they are first principles of the science. Other human sciences do not often talk in terms of laws, or at least ones that have caught on in the profession, though I can think of a possible fundamental one in psychology. Our method of thinking—our unique, individual method of using our minds, "psycho-epistemology"—that has developed since childhood affects our thought and behavior in the present.[13]

[11] See George Reisman, *Capitalism*, 618–22, for an interesting algebraic derivation, using the supply-demand formula for wages, to show that the productivity of labor is the fundamental determinant of real wages. He goes on to point out the primary variables relevant in increasing labor productivity are the capital-to-consumer goods ratio (the economic degree of capitalism) and, probably most important, technological progress (for example, backhoe versus pick and shovel).

[12] The supply-demand equation, $p = d/s$, is similar to $f = ma$ (force equals mass times acceleration) in that there are no constants in either equation. Newton's second law, however, describes physical matter and uses standard units of measurement for both mass and acceleration. Mises cautioned economists about the misleading use in their field of metaphors from Newtonian mechanics, such as friction and mechanism. *Human Action*, 114, 117, 311–12. Aristotle did not know the modern methods of mathematical measurement, though in *On the Heavens*, 297b15–298a20, in McKeon, *Basic Works*, 436–37, he correctly established that the earth was round and cited mathematicians who gave an early estimate of the earth's circumference.

[13] And that present thought and behavior may be healthy or unhealthy. Perhaps this could be called Freud's Law. The idea of the unconscious influencing the conscious predates Freud and he did not come up with the concept of psycho-epistemology. His work, nonetheless, significantly influenced subsequent psychology. See Whyte, *The Unconscious before Freud*.

Laws are principles, first or derivative. I most often use the word "principle," hence my frequent reference to "concepts and principles."

KANT'S LEGACY

Modern philosophy, from Hobbes and Descartes to Kant, as well as contemporary philosophy from the nineteenth century to today, have failed to solve the problem of universals, or even to pay much attention to it. Post-Renaissance philosophers, it seems, have spent much of their time, and ink, struggling to get outside of consciousness, in the process—and centuries—failing to find "true" reality.[14] Defense of the free society, however, as I stated in the preface to this work, requires a sound theory of universals. Plato's defective theory led to his advocacy of an authoritarian society. Thomas Hobbes said there is no objective theory, therefore the "public sword" must be used to settle disputes. Karl Marx, under the influence of Kant, preached polylogism and paved the way for what we have today: postmodernism.

Instead of saying as Marx did, "Your logic is bourgeois, you cannot understand proletarian logic," the postmoderns updated Marx to the more general "You have your truth and I have mine."[15] As philosopher Stephen Hicks writes, postmodernism is the ultimate conclusion of Kant's legacy, "a principled separation of subject and object, of reason and reality," meaning we get one or the other, not both. Postmodernism, he writes a little later, is a "ruthlessly consistent statement of the rejection of reason."[16]

Given these epistemological premises, Hicks then asks, why are the postmoderns still vehemently supporting the socialist political and economic program—despite socialism's many past and present failures? After all, their epistemology says nothing can be established as true or false. Therefore, no system can be proven to be better or worse than

---

[14] Cf. Reardan, *Emotions*, 19–25. "The history of Western philosophy, it may be said, has been a long struggle first to separate consciousness as a phenomenon in its own right, distinct from the world of which it is conscious, and then to put consciousness and the world back together again, to show the inseparable relationship between them"—as a natural, biological process, neither materialistic nor supernatural. Reardan, 19.

[15] Usually adding, "We only have narratives, not knowledge." My response is to substitute the word "fiction" for narrative. You will then more clearly understand what the postmoderns mean: "You have your *fiction*, I have mine."

[16] Stephen R. C. Hicks, *Explaining Postmodernism: Skepticism and Socialism from Rousseau to Foucault* (Scholargy Publishing, 2004), 42, 81. Italics omitted.

any other. The answer Hicks gives is that if reason and evidence are out, feelings and emotion are all that is left, and the postmoderns opted for feelings so they could retain their irrational embrace of socialism.[17]

I am tempted to add that postmodernism is not just a complacent skepticism taken from Kant and the positivists. It is a *militant* skepticism. Facts do not matter to the postmoderns, and *they are militant.* Language, which means reason, is not connected to reality, but it can be a useful tool for rhetorically defeating opponents in debates, often with name calling and elevated volume, explicit threats, and, sometimes, outright violence.[18] When language fails, violence becomes the means of resolving disputes. This is Kant's legacy.

AYN RAND'S LEGACY

Ayn Rand's legacy is the contribution of several important ideas to both philosophy and science: abstraction as measurement omission; definitions as summaries and condensations of concepts; abstraction from abstractions as hierarchical and contextual; essences, concepts, and objectivity as epistemological; truth as identification; and logic as non-contradictory identification, fundamentally of the "in here" with the "out there." Her theory of value as applying to all living organisms must be included here and, especially, as she applied it to human beings as the base of ethics and political philosophy. Ayn Rand brought Aristotle's axiom of the primacy of existence back to philosophy.

Her theory of concepts acquits Sigmund Freud and Ludwig von Mises of the charges of their many critics. Freud's psychoanalysis, for example, is said to be mere "literary psychology," and Mises's economics is non-quantitative and, today, passé. Both Freud and Mises, so the critics say, are not "real" scientists and, if we refer back to chapter 1 of this work, primatologist Jane Goodall was similarly dismissed for her anthropomorphism and, of course, her lack of credentials.[19]

Ayn Rand is criticized, or rather ignored, in this same *scientistic* manner. She is, the critics say, when they do mention her, just a novelist, not a real philosopher. And in my career, a "peer" reviewer once commented on a paper I had submitted for publication: "I don't consider Ayn Rand to be a reference for anything."

---

[17] Hicks, 89.

[18] Hicks, 176.

[19] See above, pp. 22–23.

Rand's answer to the problem of universals is what is needed today to provide the logical foundation of science—a foundation not just of science in general, but a foundation of the human sciences, especially psychology and economics, which in turn provides the intellectual defense of the free society.

It is ironic that laypersons and professionals, especially engineers who build rockets and bridges, assume that there are valid universals, valid facts, and valid truth. Professional philosophers seem to be the only ones who have a problem with the so-called problem of universals. For many years, after reading Ayn Rand's theory, my response to this "problem" has been, and continues to be, "What's the problem?"

Bibliography

Adler, Mortimer J. *The Difference of Man and the Difference It Makes.* World Publishing, 1968.

Arieti, Silvano. *Interpretation of Schizophrenia.* Basic Books, 1955. Second edition, 1974.

Aristotle. *The Basic Works of Aristotle.* Edited by Richard McKeon. Random House, 1941.

Aristotle. *The Complete Works of Aristotle: The Revised Oxford Translation.* Edited by Jonathan Barnes. 2 vols. Princeton University Press, 1984.

Bacon, Francis. *The New Organon.* Edited by Lisa Jardine and Michael Silverthorne. Cambridge University Press, 2000. Originally published in Latin in 1620.

Basharzad, Mani. "Milton Friedman Regretted Writing 'The Methodology of Positive Economics.'" *Mises Institute.* December 27, 2024. https://mises.org/power-market/milton-friedman-regretted-writing-methodology-positive-economics.

Bettelheim, Bruno. *Freud and Man's Soul.* Vintage Books, 1983.

Binswanger, Harry. *The Biological Basis of Teleological Concepts.* Ayn Rand Institute Press, 1990.

Binswanger, Harry. "Volition as Cognitive Self-Regulation." *Organizational Behavior and Human Decision Processes* 50, no. 2 (1991): 154–78. https://doi.org/10.1016/0749-5978(91)90019-P.

Branden, Nathaniel. "Intellectual Ammunition Department: Is there any validity to the claim that certain things are unknowable." *The Objectivist Newsletter* 2, no. 1 (1963): 3.

Branden, Nathaniel. *The Psychology of Self-Esteem: A Revolutionary Approach to Self-Understanding that Launched a New Era in Modern Psychology.* 1969. Reprint, Jossey-Bass, 2001.

Branden, Nathaniel. "The Stolen Concept," *The Objectivist Newsletter* 2, no. 1 (1963): 2, 4.

Bregman, Rutger. *Humankind: A Hopeful History.* Little, Brown, 2019.

Brentano, Franz. *Psychology from an Empirical Standpoint.* Translated by Antos C. Rancurello, D. B. Terrell, and Linda L. McAlister. Routledge, 1973. First published in German in 1874.

Bostaph, Samuel. "The Methodological Debate Between Carl Menger and the German Historicists." *Atlantic Economic Journal* 6, no. 3 (1978): 3–16. https://doi.org/10.1007/BF02313305.

Breuer, Josef. "Fräulein Anna O." In Breuer and Freud, *Studies on Hysteria.*

Breuer, Josef and Sigmund Freud, *Studies on Hysteria.* Translated and edited by James Strachey. Basic Books, 1955. First published in German in 1895.

Brull, H. Frank. "A Reconsideration of Some Translations of Sigmund Freud," *Psychotherapy: Theory, Research & Practice* 12, no. 3 (1975): 273–79.

Childs, Jr., Roy A. "Big Business and the Rise of American Statism." In *Liberty Against Power: Essays by Roy A. Childs., Jr.* Edited by Joan Kennedy Taylor. Fox & Wilkes, 1994.

Damasio, Antonio. *Feeling and Knowing: Making Minds Conscious.* Vintage Books, 2021.

Doidge, Norman. *The Brain that Changes Itself.* Penguin, 2007.

Domarus, Eilhard von. "The Specific Laws of Thought in Schizophrenia." In *Language and Thought in Schizophrenia.* Edited by J. S. Kasanin. University of California Press, 1944.

Doyle, Arthur Conan. *A Study in Scarlet: A Detective Story.* 1888. Reprint, SeaWolf Press, 2018

Ducheyne, Steffen, and John P. McCaskey. "The Sources of Mill's Views of Ratiocination and Induction." In *Mill's A System of Logic: Critical Appraisals.* Edited by Antis Loizides. Routledge, 2014.

Ferenczi, Sándor, Karl Abraham, Sigmund Freud, Ernest Jones, and Ernst Simmel. *Psycho-Analysis and the War Neuroses.* Edited by Ernest Jones. International Psycho-Analytical Press. 1921. The Project Gutenberg eBook of Psycho-Analysis and the War Neuroses. https://www.gutenberg.org/ebooks/55443.

Freud, Anna. *The Ego and the Mechanisms of Defence.* Translated by Cecil Baines. International Universities Press, 1946. First published in German in 1936.

Freud, Sigmund. "Analysis of a Phobia in a Five-Year Old Boy (Little Hans)." In *Case Histories: Dora and Little Hans.* Translated and edited by Alix and James Strachey. 1962. Present volume edited by Angela Richards. Penguin Books, 1977. First published in German in 1909.

Freud, Sigmund. *Beyond the Pleasure Principle.* In *Beyond the Pleasure Principle and Other Writings.* Translated by John Reddick. Penguin, 2003. First published in German in 1920.

Freud, Sigmund. *Civilization and Its Discontents.* Translated by Ulrich Baer. Warbler Press, 2022. First published in German in 1930.

Freud, Sigmund. "A Difficulty in the Path of Psycho-Analysis." Translated by Joan Riviere. In Sigmund Freud, *Psychological Writings and Letters.* Edited by Sandor L. Gilman. Continuum Publishing, 1995. First published as a journal article in Hungarian and German in 1917.

Freud, Sigmund. *The Ego and the Id.* Translated by Joan Riviere. Revised and edited by James Strachey. 1960. Reprint, Norton, 1989. First published in German in 1923.

Freud, Sigmund. *Five Lectures on Psycho-Analysis.* Translated by James Strachey. Norton, 1961. First published in German in 1910.

Freud, Sigmund. *Inhibition, Symptom, and Fear.* In *Beyond the Pleasure Principle and Other Writings.* Translated by John Reddick. Penguin, 2003. First published in German in 1926.

Freud, Sigmund. *The Interpretation of Dreams.* Translated and edited by James Strachey. 1955. Reprint, Basic Books, 2010. First published in German in 1900.

Freud, Sigmund. *Introductory Lectures on Psycho-Analysis.* Translated and edited by James Strachey. Norton, 1963. Reprint, 1989. First published in German 1916–17.

Freud, Sigmund. *The Letters of Sigmund Freud and Arnold Zweig.* Edited by Ernst L. Freud. Translated by Elaine and William Robson-Scott. Harcourt Brace, 1970.

Freud, Sigmund. *The Letters of Sigmund Freud to Eduard Silberstein 1871–1881.* Edited by Walter Boehlich. Translated by Arnold J. Pomerans. Belknap Press, 1990.

Freud, Sigmund. *Mass Psychology and Analysis of the I.* In *Mass Psychology and Other Writings.* Translated by J. A. Underwood. Penguin, 2004. First published in German in 1921.

Freud, Sigmund. *New Introductory Lectures on Psycho-Analysis.* Translated and Edited by James Strachey. Norton, 1964. Reprint, 1989. First published in German in 1933.

Freud, Sigmund. *On Dreams.* Translated and edited by James Strachey. Norton, 1952. Reprint, 1989. First published in German in 1901.

Freud, Sigmund. *On the Introduction of Narcissism.* In *Beyond the Pleasure Principle and Other Writings.* Translated by John Reddick. Penguin, 2003. First published in German in 1914.

Freud, Sigmund. *An Outline of Psychoanalysis.* Translated by Helena Ragg-Kirby. In *The Penguin Freud Reader.* Penguin, 2006. Edited by Adam Phillips. First published in German posthumously in 1940.

Freud, Sigmund. "Project for a Scientific Psychology." In the *Standard Edition: Pre-Psychoanalytic Publications and Unpublished Drafts.* Vol. 1 (1866–1899). Translated by James Strachey. 1954. Reprint, Vintage Books, 2001. Written in 1895, published in German in 1950 and English in 1954.

Freud, Sigmund. *The Psychopathology of Everyday Life.* Translated and edited by James Strachey. Norton, 1960. First published in German in 1901.

Freud, Sigmund. "Psychotherapy of Hysteria." In Breuer and Freud, *Studies on Hysteria.*

Freud, Sigmund. *The Question of Lay Analysis.* 1950. Reprint, Norton, 1970. First published in German in 1926.

Freud, Sigmund. "Remembering, Repeating, and Working Through." In *Beyond the Pleasure Principle and Other Writings.* Translated by John Reddick. Penguin, 2003. First published in German as a journal article in 1914.

Freud, Sigmund. "Sigmund Freud on Psychoanalysis." *Encyclopedia Britannica.* November 6, 2014. https://www.britannica.com/topic/Sigmund-Freud-on-psychoanalysis-1983319. First published in English in 1926.

Freud, Sigmund. *The Standard Edition of The Complete Psychological Works of Sigmund Freud.* Translated and edited by James Strachey, 24 vols. Hogarth Press, 1953–74.

Freud, Sigmund. *Three Essays on the Theory of Sexuality.* Translated and edited by James Strachey (Basic Books, 1962. First published in German in 1905.

"Freud's Method for Interpreting Dreams," *Freud Museum London.* https://www.freud.org.uk/schools/resources/the-interpretation-of-dreams/freuds-method-for-interpreting-dreams/.

Friedman, Milton. "The Methodology of Positive Economics." In *Essays in Positive Economics.* University of Chicago Press, 1953.

Gay, Peter. *Freud: A Life for Our Time.* Norton, 1988.

Glasser, William. *Choice Theory: A New Psychology of Personal Freedom.* Harper, 1998.

Goodall, Jane. *In the Shadow of Man.* 1971. Reprint, Mariner Books, 2000.

Gordon, David. *The Philosophical Origins of Austrian Economics.* Mises Institute, 1993.

Greaves, Bettina Bien, and Robert W. McGee. *Mises: An Annotated Bibliography.* Foundation for Economic Education, 1993.

Gross, Gail. "How Your Baby Learns Language in the Womb," *Psychology Today,* September 28, 2022. https://www.psychologytoday.com/us/blog/strategies-living/202209/how-your-baby-learns-language-in-the-womb.

Harré, Rom. *Great Scientific Experiments: Twenty Experiments that Changed Our View of the World.* 1981. Reprint, Dover Publications, 2002.

Hayek, F. A. "The Pretence of Knowledge." *Friedrich von Hayek Prize Lecture.* December 11, 1974. https://www.nobelprize.org/prizes/economic-sciences/1974/hayek/lecture/.

Hayek, F. A. "The Use of Knowledge in Society." In *Individualism and Economic Order.* Gateway Editions, 1948.

Hicks, Stephen R. C. *Explaining Postmodernism: Skepticism and Socialism from Rousseau to Foucault.* Scholargy Publishing, 2004.

Hülsmann, Jörg Guido. Introduction to the Third Edition of *Epistemological Problems of Economics,* by Ludwig von Mises, 3rd ed. Translated by George Reisman. Mises Institute. (2003).

Hülsmann, Jörg Guido. *Mises: The Last Knight of Liberalism.* Mises Institute, 2007.

Ioannidis. John P. A. "Why Most Clinical Research Is Not Useful." *PloS Medicine* 13, no. 6 (2016): 1–10, e1002049. https://doi.org/10.1371/journal.pmed.1002049.

Ioannidis, John P. A. "Why Most Published Research Findings Are False," *PLoS Medicine* 2, no. 8 (2005): 696–701. https://doi.org/10.1371/journal.pmed.0020124.

Joseph, H. W. B. *An Introduction to Logic.* Oxford University Press, 1916.

Kant, Immanuel. *Critique of Pure Reason.* Translated and edited by Paul Guyer and Allen W. Wood. Cambridge University Press, 1998. First published in German in 1781, revised in 1787.

Kant, Immanuel. *Groundwork of the Metaphysics of Morals.* Translated by Mary Gregor. Cambridge University Press, 1998. First published in German in 1785.

Kant, Immanuel. *Prolegomena to Any Future Metaphysics.* Edited by Lewis White Beck. Liberal Arts Press, 1950. First published in German in 1783.

Kaufmann, Walter. *Discovering the Mind.* Vol. 1, *Goethe, Kant, and Hegel.* Vol. 2, *Nietzsche, Heidegger, and Buber.* Vol. 3, *Freud versus Adler and Jung.* McGraw-Hill, 1980.

Kelley, David. *The Art of Reasoning.* Norton, 1988.

Kelley, David. "Concepts and Propositions." *The Atlas Society.* June 6, 2025. https://www.atlassociety.org/concepts-and-propositions.

Kemp Smith, Norman. *A Commentary to Kant's "Critique of Pure Reason."* Macmillan, 1918.

Kim, Hojin I. and Scott P. Johnson. "Infant Perception." In *Encyclopedia Britannica.* May 28, 2014. https://www.britannica.com/topic/infant-perception.

Kirkpatrick, Jerry. *Applying Principles: Short Essays Based on the Philosophy of Ayn Rand, Economics of Ludwig von Mises, and Psychology of Edith Packer.* Kirkpatrick Books, 2021.

Kirkpatrick, Jerry. "Chance Emotional Generalizations and Introspection." *Jerry Kirkpatrick's Blog.* June 7, 2019. https://jerrykirkpatrick.blogspot.com/2019/06/chance-emotional-generalizations-and.html.

Kirkpatrick, Jerry. *In Defense of Advertising: Arguments from Reason, Ethical Egoism, and Laissez-Faire Capitalism.* 1994. Paperback edition, TLJ Books, 2007.

Kirkpatrick, Jerry. *Independent Judgment and Introspection: Fundamental Requirements of the Free Society.* Kirkpatrick Books, 2019.

Kirkpatrick, Jerry. *Montessori, Dewey, and Capitalism: Educational Theory for a Free Market in Education.* Kirkpatrick Books, 2008.

Kirkpatrick, Jerry. "Objectivist Epistemology as the Foundation of Marketing Theory." In *Perspectives on Ayn Rand's Contributions to Economic and Business Thought.* Edited by Edward W. Younkins. Rowman & Littlefield, 2018.

Kirkpatrick, Jerry. "On the Nature of Human Nature," *Jerry Kirkpatrick's Blog,* July 22, 2025, https://jerrykirkpatrick.blogspot.com/2025/07/on-nature-of-human-nature.html.

Kirkpatrick, Jerry. "Reisman's Net Consumption, Net Investment Theory of Aggregate Profit: Exposition and Comparison to Mises and Rothbard." *The American Journal of Economics and Sociology* 63, no. 3 (2004): 627–46. https://www.jstor.org/stable/3488146.

Kirkpatrick, Jerry. "Theory and History in Marketing." *Managerial and Decision Economics* 4, no. 1 (1983): 44–49. https://doi.org/10.1002/mde.4090040106.

Kirsner, Douglas. "Fresh Freud: No Longer Lost in Translation." *Psychoanalytic Psychology* 24, no. 4 (2007): 658–66. https://doi.org/10.1037/0736-9735.24.4.658.

Klein, Richard G. "A Diary or Chronology of Important Events in Freud's Life," *Freud 2 Lacan.* https://freud2lacan.b-cdn.net/freud_diary.pdf.

Leroi, Armand Marie. *The Lagoon: How Aristotle Invented Science.* Penguin, 2015.

Lesher, James. "A Note on the Simile of the Rout in the *Posterior Analytics* ii 19." *Ancient Philosophy* 31, no. 1 (2011): 121-25.

Lewin, Kurt. *Principles of Topological Psychology.* McGraw-Hill, 1936.

Locke, John. *An Essay Concerning Human Understanding.* Abridged and edited by A. D. Woozley. Meridian, 1974. Original, complete edition published in 1690.

London, Jack. *The Iron Heel.* 1908. Reprint, Penguin Books, 2006.

Long, Roderick. "Praxeology: Who Needs It." *The Journal of Ayn Rand Studies* 6, no. 2 (2005): 299–316.

Maslow, Abraham. *Toward a Psychology of Being.* 1962. Reprint, Wilder Publications, 2011.

Mayhew, Robert, ed. *Ayn Rand's Marginalia: Her Critical Comments on the Writings of over 20 Authors.* Second Renaissance Press, 1995.

McCaskey, John P. "The History of Induction," *John P. McCaskey* (blog). March 12, 2015. https://www.johnmccaskey.com/history-of-induction/.

McCaskey, John P. "History of 'Temperature': Maturation of a Measurement Concept," *Annals of Science* 77, no. 4 (2020): 399–444. https://doi.org/10.1080/00033790.2020.1817980.

McCaskey, John P. "Induction, Philosophical Conceptions of." In *Encyclopedia of Renaissance Philosophy.* Edited by M. Sgarbi. Springer, 2020. https://doi.org/10.1007/978-3-319-02848-4_1054-2.

McCaskey, John P. "Induction in the Socratic Tradition." In *Shifting the Paradigm: Alternative Perspectives on Induction.* Edited by Louis F. Groarke & Paolo C. Biondi. De Gruyter, 2014. https://doi.org/10.1515/9783110347777.toc.

McCaskey, John P. "Induction without the Uniformity Principle." *John P. McCaskey* (blog). May 12, 2015. https://www.johnmccaskey.com/uniformity-principle/.

McCaskey, John P. "Key to Induction: Distinguish General and Universal." *John P. McCaskey* (blog). June 11, 2014. https://www.johnmccaskey.com/generalizations/.

McCloskey, Deirdre Nansen. *Beyond Positivism, Behaviorism, and Neoinstitutionalism in Economics.* University of Chicago Press, 2022.

Mellor, Mark. "Elusive Objects of Translation," *Cadenza Academic Translations.* https://www.cadenza-academictranslations.com/papers-on-translation/.

Menger, Carl. *Principles of Economics.* Translated by James Dingwall and Bert F. Hoselitz. 1976. Reprint, New York University Press, 1981. First published in German in 1871.

Mises, Ludwig von. *The Anti-Capitalistic Mentality.* 1956, published in English. Reprint, Mises Institute, 2008.

Mises, Ludwig von. *Bureaucracy.* 1944, published in English. Reprint, Mises Institute, 2008.

Mises, Ludwig von. *Critique of Interventionism.* Translated by Hans F. Sennholz. 1976. Reprint, Mises Institute, 2011. First published in German in 1929.

Mises, Ludwig von. *Economic Calculation in the Socialist Commonwealth.* Translated by S. Adler. Mises Institute, 1990. Published as a journal article in German in 1920 and translated into English in 1935.

Mises, Ludwig von. *Epistemological Problems of Economics*. 3rd ed. Translated by George Reisman. 1960. Reprint, Mises Institute. (2003). First published in German in 1933.

Mises, Ludwig von. *Human Action: A Treatise on Economics*. 1949. Reprinted as scholar's edition, Mises Institute, 1998. First published in German as *Nationalökonomie* in 1940, revised and edited by the author for the 1949 English edition.

Mises, Ludwig von. *Liberalism in the Classical Tradition*. Translated by Ralph Raico, 3rd ed. Foundation for Economic Education and Cobden Press, 1985. First published in German in 1927.

Mises, Ludwig von. *Memoirs*. Translated by Arlene Oost-Zinner. Mises Institute, 2009. Written in the early 1940s but not published until 1978 in both German and English.

Mises, Ludwig von. *Nation, State, and Economy: Contributions to the Politics and History of Our Time*. Translated by Leland B. Yeager. New York University Press, 1983. First published in German in 1919.

Mises, Ludwig von. *Omnipotent Government*. 1944, published in English. Reprint, Mises Institute, 2010.

Mises, Ludwig von. *Planned Chaos*. 1947, published in English. Reprint, Mises Institute, 2014.

Mises, Ludwig von. *Socialism: An Economic and Sociological Analysis*. Translated by J. Kahane. 1936. Reprint, Yale University Press, 1951. First published in German in 1922.

Mises, Ludwig von. *Theory and History: An Interpretation of Social and Economic Evolution*. 1957, published in English. Reprint, Mises Institute, 2007.

Mises, Ludwig von. *The Theory of Money and Credit*. Translated by H. E. Batson. 1934. Reprint, Liberty Classics, 1981. First published in German in 1912.

Mises, Ludwig von. *Ultimate Foundation of Economic Science*, published in English. D. Van Nostrand, 1962.

Packer, Edith. *Lectures on Psychology: A Guide to Understanding Your Emotions*. TJS Books, 2012. Kindle. Paperback edition, 2018.

Peikoff, Leonard. "The Analytic-Synthetic Dichotomy." In Ayn Rand, *Introduction to Objectivist Epistemology*. Expanded 2nd ed. NAL Books, 1990. First published in *The Objectivist* 6, nos. 6–9 (1967).

Peikoff, Leonard. *Objectivism: The Philosophy of Ayn Rand*. Dutton, 1991.

Peikoff, Leonard. *The Ominous Parallels: The End of Freedom in America*. Stein and Day, 1982.

Popper, Karl. *Conjectures and Refutations: The Growth of Scientific Knowledge*. Routledge and Kegan Paul, 1963.

"Positivism." In *The Blackwell Dictionary of Twentieth-Century Social Thought*. Blackwell, 1993.

Posner, Richard. *The Little Book of Plagiarism*. Pantheon, 2007.

Rand, Ayn. *The Art of Non-Fiction: A Guide for Writers and Readers*. Edited by Robert Mayhew. Penguin, 2001.

Rand, Ayn. *Atlas Shrugged*. Random House, 1957.

Rand, Ayn. "Don't Let It Go." In *Philosophy: Who Needs It.* Bobbs-Merrill, 1982. First published in *The Ayn Rand Letter* 1, nos. 4–5 (1971).

Rand, Ayn. "For the New Intellectual." In *For the New Intellectual: The Philosophy of Ayn Rand.* Signet Books, 1961.

Rand, Ayn. *Introduction to Objectivist Epistemology.* NAL Books, 1990. First published in *The Objectivist* 5, nos. 5–12; 6, nos. 1–2 (1966–67). Expanded 2nd ed., with "The Analytical-Synthetic Dichotomy," by Leonard Peikoff, and excerpts from epistemology workshops, 1969–71.

Rand, Ayn. "The Metaphysical Versus the Man-Made," in *Philosophy: Who Needs It*, Bobbs-Merrill, 1982. First published in *The Ayn Rand Letter* 2, nos. 12–13 (1973).

Rand, Ayn. "The Objectivist Ethics." In *The Virtue of Selfishness: A New Concept of Egoism.* NAL Books, 1964. Originally a lecture delivered February 9, 1961, at the University of Wisconsin, Madison.

Rand, Ayn. "Philosophical Detection." In *Philosophy: Who Needs It.* Bobbs-Merrill, 1982. First published in *The Ayn Rand Letter* 3, nos. 9–10 (1974).

Rand, Ayn. "A Statement of Policy." *The Objectivist* 7, no. 6 (1968): 7–9.

Rand, Ayn. "What Is Capitalism?" In *Capitalism: The Unknown Ideal.* New American Library, 1966. First published in The Objectivist Newsletter 4, nos. 11–12 (1965).

Reardan, Linda. *Emotions: A Modern Aristotelian View.* Kirkpatrick Books, 2026.

Reisman, George. *Capitalism: A Treatise on Economics.* Jameson Books, 1996.

Rickert, Heinrich. *The Limits of Concept Formation in Natural Science: A Logical Introduction to the Historical Sciences.* Edited and translated by Guy Oakes. Abridged edition. Cambridge University Press, 1986. First published in German in 1895.

Rickert, Heinrich. *Science and History: A Critique of Positivist Epistemology.* Translated by George Reisman. Edited by Arthur Goddard. D. Van Nostrand, 1962. First published in German in 1899.

Rizzo, Francis. "Memoirs of an Invisible Man." *Opera News.* February 5, 12, 19, and 26, 1972. https://freud2lacan.b-cdn.net/Interview_with_Little_Hans.pdf.

Rothbard, Murray. *The Essential Ludwig von Mises.* 1973, 1988. Reprint, Mises Institute, 2009.

Samenow, Stanton. *Inside the Criminal Mind*, 1984. Revised edition, Broadway Books, 2014.

Searle, John. *Minds, Brains, and Science.* Harvard University Press, 1984.

Shedler, Jonathan. "It's Time for Psychology to Lead, Not Follow." *Psychology Today.* Updated February 25, 2025. https://www.psychologytoday.com/us/blog/psychologically-minded/201910/its-time-for-psychology-to-lead-not-follow.

Shedler, Jonathan. "Selling Bad Therapy to Trauma Victims," *Psychology Today.* November 19, 2017. https://www.psychologytoday.com/us/blog/psychologically-minded/201711/selling-bad-therapy-trauma-victims.

Shedler, Jonathan. "That Was Then, This Is Now: Psychoanalytic Psychotherapy for the Rest of Us." *Contemporary Psychoanalysis* 58, nos. 2–3 (2022): 405–37. https://doi.org/10.1080/00107530.2022.2149038.

"Sigmund Freud's Famous Psychoanalytic Couch." *Freud Museum London.* https://www.freud.org.uk/about-us/the-house/sigmund-freuds-famous-psychoanalytic-couch/.

Smith, Adam. *An Inquiry in the Nature and Causes of the Wealth of Nations.* 2 vols. in one. Edited by Edwin Cannan. 1904. Reprint, University Chicago Press, 1976. Original edition published in 1776.

Smith, Barry. *Austrian Philosophy: The Legacy of Franz Brentano.* Open Court, 1994.

Snyder, Laura J. *The Philosophical Breakfast Club.* Broadway Books, 2011.

Solms, Mark. "Notes on the Revised Standard Edition," *Psychoanalytic Review* 100, no. 1 (2013): 201–210. https://doi.org/10.1521/prev.2013.100.1.20.

Stevens, S. S. "On the Theory of Scales of Measurement." *Science* 103, no. 2684 (1946), 677–80, https://www.jstor.org/stable/1671815.

Stewart, Sophia. "Freud's Writings Get an Update—30 Years in the Making." *Publisher's Weekly.* August 1, 2024.

Szasz, Thomas. *Ethics of Psychoanalysis: The Theory and Method of Autonomous Psychotherapy.* Syracuse University Press, 1965.

Taieb, Hamid. "Intentionality and Reference: A Brentanian Distinction." *The Monist* 100, no. 1 (2017): 100–32. https://doi.org/10.1093/monist/onw022.

Whitaker, Robert. *Mad in America: Bad Science, Bad Medicine, and the Enduring Mistreatment of the Mentally Ill.* Basic Books, 2002.

Whyte, Lancelot Law. *The Unconscious before Freud: A History of the Evolution of Human Awareness.* Basic Books, 1960.

Yochelson, Samuel, and Stanton Samenow. *The Criminal Personality.* Vol. 1, *A Profile for Change.* 1976. Reprint, Jason Aronson, 1993.

Index

Cover by 1106Design, 1106design.com

Interior design by Kirkpatrick Books

Main text: Warnock Pro

Headings: Myriad Pro

www.ingramcontent.com/pod-product-compliance
Lightning Source LLC
LaVergne TN
LVHW101502200926
848483LV00001B/204
* 9 7 8 1 7 3 7 1 3 8 7 5 4 *